OZARK WILDFLOWERS

Ozark Wildflowers

Thomas E. Hemmerly

THE UNIVERSITY OF GEORGIA PRESS

ATHENS AND LONDON

Athens, Georgia 30602

Designed by Erin Kirk New
Set in Berkeley Old Style Medium by G&S Typesetters
Printed and bound through Pacifica Communications

The paper in this book meets the guidelines for
permanence and durability of the Committee on
Production Guidelines for Book Longevity of the
Council on Library Resources.

Printed in Korea
02 03 04 05 06 C 5 4 3 2 1
02 03 04 05 06 P 5 4 3 2 1

Library of Congress Cataloging-in-Publication Data
Hemmerly, Thomas E. (Thomas Ellsworth), 1932–
Ozark wildflowers / Thomas E. Hemmerly.
p. cm.
Includes bibliographical references (p.) and index.
ISBN 0-8203-2336-5 (alk. paper)—ISBN 0-8203-2337-3 (pbk. : alk. paper)
1. Wild flowers—Ozark Mountains Region—Identification. 2. Wild flowers—
Ozark Mountains Region—Pictorial works. I. Title.
QK134 .H46 2002
582.13'09767'1—dc21 2001047648

British Library Cataloging-in-Publication Data available

To my son, Everett,

and my daughter, Kathy Roadarmel

"A man's life does not consist in the abundance of his possessions."

—Luke 12:15

Contents

PINK

BLUE/PURPLE

GREEN/BROWN

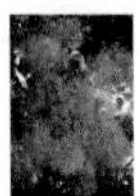

Author's Notes

The Ozark country, a place of nostalgia, legend, and mystery, has long held a special fascination for me. As a boy, I relished the small-town antics of the radio duo Lum and Abner of the Jot-em-down Store of Pine Bluff, Arkansas. More recently, I have delighted in the outlandish, sometimes bawdy, tales collected by the folklorist Vance Randolph; also in the poignant, compelling, though moralistic, stories of Harold Bell Wright in such classic books as his acclaimed *Shepherd of the Hills.*

From these and other reflections of earlier times in the Ozarks, one is impressed with an image of a simple, backward folk living in a time warp. If such a description of Ozark people were once true, it no longer is. Within the area are thriving cities as well as prosperous ranches, farms, and vast, protected natural areas that sustain an astonishing diversity of plants and animals.

My interest in the natural history of the Ozarks began with family camping trips in the 1960s. Included in our itinerary were such places as Bull Shoals, Crowley's Ridge, and Petit Jean State Park. Though its boundaries are indistinct, the core Ozark region is a large (approximately 50,000-square-mile) upland area that sprawls across at least four states. In addition to the Ozarks proper, this book also covers the Ouachita Mountains (22,000 square miles) to the south, and Crowley's Ridge, a narrow upland that extends from eastern Arkansas into eastern Missouri.

Because of its varied topography and its location at the crossroads of the southern Midwest, this Ozark region has a rich and distinctive biota. Among its plants are those characteristic of prairies to the north and west, of southern coastal plains to the south, and of the Appalachians. Also there are many species that are essentially Ozarkian in distribution.

Using this book you should be able not only to identify most flowering herbs, shrubs, and trees of the Ozarks that you encounter, but also to appreciate them in the context of the ecosystems of which they are a part. Furthermore, medical and other ethnobotanical information is provided, indicating the uses of plants by Native Americans and early European settlers as well as modern uses.

The inevitable accelerated commercial development of the Ozarks in the twenty-first century gives added importance to publications that document and interpret the native plants and animals of this special region. Knowledge leads to appreciation, and appreciation should lead to the desire to protect Ozark plants, associated animals and other organisms, and the habitats that sustain them.

Acknowledgments

As is the case with other American natural history books, *Ozark Wildflowers* draws upon centuries of exploration and discoveries by numerous early naturalists. Following preliminary botanical studies of eighteenth-century workers, modern studies of the Ozark flora were initiated in the nineteenth century by notable botanists such as George Engelmann (1809–84) and Heinrich K. D. Eggert (1841–1904), and others. Many were physicians educated in Germany who later became associated with the Missouri Botanical Garden in St. Louis. Few, if any, would be characterized as "Ozark botanists," as they generally collected over much of the central United States.

Among the twentieth (and twenty-first) century botanists on whom I have relied are the following: Julian Steyermark (1909–88), *Flora of Missouri*, 1963; Edgar W. Denison (1904–93), *Missouri Wildflowers*, 1998; Robert H. Mohlenbrock (1931–), *Flora of Southern Illinois*, with John Voigt, 1959; Carl G. Hunter (1923–), *Wildflowers of Arkansas*, 1988, and *Trees, Shrubs and Vines of Arkansas*, 1989; Edwin B. Smith (1936–), *An Atlas and Annotated List of the Vascular Plants of Arkansas*, 1988; George A. Yatskievych (1957–), *Steyermark's Flora of Missouri*, volume 1, 1999. To all of these, I am indebted.

Earl Hendrix, of the Sylamore District, Ozark National Forest, Mountain View, Arkansas, provided useful field assistance. The skillful typing of the manuscript was performed primarily by Emily Kee; others involved were Cody Smith, Tiffany Freeze, Ashley Messick, and Carolyn Gray.

All photographs were made by me except for *Verbena canadensis*, by Jim Lea. The expense of printing the color plates was subsidized by a grant from Catherine Keever, a prominent plant ecologist and professor emerita, Millersville State University.

The staff of the University of Georgia Press deserves credit for bringing this volume to fruition. Included are Barbara Ras, senior editor, whose advice and guidance is well appreciated; Betty McDaniel, production coordinator; Ellen Harris, freelance copyeditor; and Erin Kirk New, designer. Kristine Blakeslee and Jennifer Comeau served as project editors. Also to be recognized are two anonymous reviewers whose comments improved the text.

Using This Book

Rather than being viewed as isolated objects, wildflowers are best considered in the context of their environment. Included are other associated plants and animals as well as a plethora of soil, water, and climatic factors, all of which, separately and collectively, exert their influences. Therefore, this book begins with a brief overview of the Ozarks (chap. 1)—its regions, geology, soils, and climate. Following is a brief description of its principal ecosystems (chaps. 2 and 3). Chapter 4 reviews the scientific naming system and variations in leaves, flowers, and other parts used in plant identification. This introduction to Ozark ecology and terminology serves as a background for Part 2.

The wildflowers illustrated in this book are arranged primarily by color and secondarily by the grand group to which they belong: monocot or dicot. Further explanation of this system is given in "Using the Color Plates."

The task of choosing which of the estimated approximately 2,000 Ozarkian species to feature was a difficult one. In general, an attempt was made to represent the considerable plant diversity of the Ozark region. Wildflowers that are common and showy are more likely to be featured than uncommon, inconspicuous ones. In spite of their ecological importance, grasses and their relatives, rushes and sedges, are not included. Trees, shrubs, and woody vines, especially those with showy flowers, are included, although they are not as well represented as are herbaceous (nonwoody) plants. Several books that provide a more thorough (and more technical) coverage of Ozark wildflowers are listed in the bibliography.

The information opposite the photograph of each species of wildflower follows a reasonably consistent pattern. The common name and scientific name are followed by the family name. Nomenclature is taken from several sources but especially Yatshievych and Turner's *Catalogue of the Flora of Missouri* and *Keys to the Flora of Arkansas* by Edwin B. Smith. If a species is known by a synonym (previously used scientific name), that name appears in brackets.

The narrative contains a brief description of the plant, its estimated relative abundance, habitat, and geographical distribution. The usual flowering time is also given. Keep in mind that flowering times may vary from place to place and year to year. When applicable, information is provided on ethnobotanical, economic, or medicinal uses. The latter is given for academic and historical purposes only; *self-medication is not recommended.* One or more similar or related species may also be described briefly, greatly increasing the coverage of the book. Less common terms used are in the glossary (appendix 1).

To identify an unknown plant, you may wish to take one of these approaches:

1. If you think you know the name, scientific or common, use the index to locate the illustration to confirm the identification. It may be the one you thought or a related one.
2. If you don't have a clue, consult the appropriate color group in the table of contents and check that section of illustrations and descriptions. Note the following explanations concerning the color sections: White includes cream and other pale tints. Red/orange includes rose and maroon; pink includes coral and mauve; blue/purple includes lavender, lilac, magenta, and violet; and green/brown includes other dark or indistinct colors as well as some flowers that are yellowish green.
3. If the plant is found in a special habitat (e.g., oak-hickory forest, cedar glade, or wetland), see the list of plants for that habitat (chaps. 2 and 3).

Many opportunities exist throughout the Ozarks for "botanizing." Some of the recommended areas are described in appendix 2.

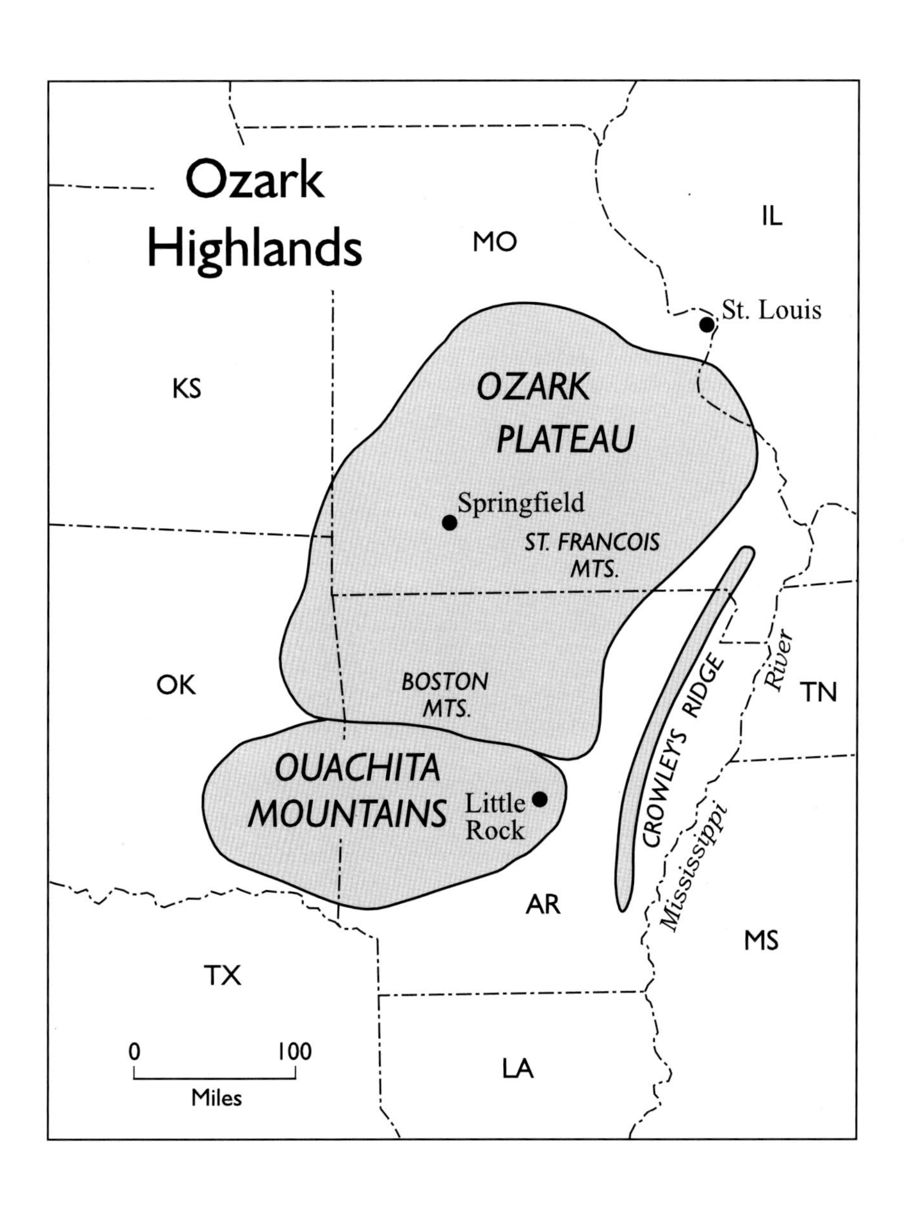

Ozark Highlands
MO
IL
St. Louis
KS
OZARK PLATEAU
Springfield
ST. FRANCOIS MTS.
OK
BOSTON MTS.
River
TN
CROWLEY'S RIDGE
OUACHITA MOUNTAINS
Little Rock
Mississippi
AR
MS
TX
0
100
Miles
LA

Part 1

Ecology of the Ozarks

CHAPTER 1

The Ozark Region

The region commonly known as the Ozarks or, by geographers, the Interior Highlands, is the most prominent landscape feature between the Rocky and Appalachian Mountains. Nevertheless, the delineation of its boundaries and the designation of its subregions are not universally agreed on by geologists, geographers, or botanists.

For our purposes, I shall use the term Ozark Highlands, or simply Ozarks, as the umbrella term for the entire region (see map). The two principal regions are the Ozark Plateau and the Ouachita Mountains. The former occupies nearly the southern half of Missouri, the northern one-fourth of Arkansas, and part of northeast Oklahoma (and a tiny portion of extreme southeast Kansas). The Ouachitas, to the south of the Ozark Plateau, extend from central Arkansas westward into eastern Oklahoma. Two smaller subregions are the Illinois Ozarks of southwestern Illinois, and Crowley's Ridge, which extends from eastern Arkansas northward into southeastern Missouri.

Birth of the Ozarks

One can hardly begin to understand the Ozarks without reference to the Appalachians. Geologists consider the Ozarks, especially the Ouachita subregion, a westward extension of the southern Appalachians. Beginning early in the North American continent's almost 4-billion-year history, the Ozarks and Appalachians were formed by the same mountain-building episodes, or orogenies. Among the events involved were continental collisions, faulting, volcanic action, and intrusions. The results of such ancient events have, of course, been greatly modified over millennia by gradual, less dramatic, yet important processes such as weathering and erosion.

Whereas adjacent regions have been repeatedly covered by seas or glaciers, the Ozarks (with the exception of the northern portion of the Illinois Ozarks) have remained high and dry for more than 250 million years. The southernmost extent of glaciation generally coincides with the present-day Missouri River. Thus, nearly the entire Ozark region has been available for colonization and habitation during this time by thousands of plant and animal species. New species have evolved while others have, no doubt, become extinct. At the same time, the nature and extent of the various plant communities have constantly changed. Such long-term events, which continue today, are further considered in the following chapter.

Subregions of the Ozarks

Though ancient they are, the Ozark Highlands today bear testimony to their origins. The Ozark Plateau remains essentially a high, flat tableland with elevations mostly between 1,500 and 2,400 feet. Its moderate relief is most noticeable in its St. Francois Mountains of southeastern Missouri and the Boston Mountains of northwestern Arkansas. The oldest portion of the Ozarks is the St. Francois Mountains, where volcanic rocks have been dated at 2 to 3 billion years old.

Just across the Mississippi River from the St. Francois Mountains are the Illinois Ozarks. Once continuous with the Ozark Plateau, this long, narrow (½- to 10-mile-wide) strip of land has been separated by the river for millions of years. To a considerable degree, however, it has retained its basic geology and its largely Ozarkian plant life.

In contrast, the Ouachita Mountains to the south of the Ozark Plateau, formed by more violent geological forces, exhibit a feature unique among mountains of the United States. They consist of ridges separated by valleys oriented east to west (both the Rockies and Appalachians extend north to

The Ouachita Mountains of eastern Oklahoma, seen from Queen Wilhelminia State Park, western Arkansas.

Vista from Mt. Magazine (highest point of the Ozarks) of the Ouachita Mountains of central Arkansas; early fall, oak-hickory forest below.

south). The Arkansas River Valley is considered the northern boundary of the Ouachitas. Hovering over the valley is Mt. Magazine; at 2,753 feet in elevation, it is the highest point of the Ozarks (and of the Midwest).

Crowley's Ridge is a narrow (½- to 12-mile-wide) sliver of elevated land within the delta region of eastern Arkansas and Missouri. According to geologists, the ridge was formed by the ancient Mississippi and Ohio Rivers, which eroded away the land on each side: the Mississippi on the west side and the Ohio on the east. Today, it extends from near Helena, Arkansas, near the Mississippi River, some 250 miles, generally paralleling the river northward, to near Commerce, Missouri, west of the state's "Bootheel." Its elevations, up to 550 feet, are well above those of the delta. Some geographers consider Crowley's Ridge to be a part of the Mississippi Lowlands instead of the Ozarks.

Rock-Soil-Plant Connections

The nature of soil is greatly influenced by the type of bedrock from which it develops. In turn, the nature of the soil determines to a considerable degree the plants that can occupy that site. This phenomenon also works in reverse, however: as plants decompose, they modify the soil. Thus there exists a reciprocal relationship between soils and plants; both are affected by the underlying rock.

Mature, residual Ozark soils, those that have formed in place over long periods of time and are composed of several layers, or horizons, are said to be stratified. At the surface is the organic or O-horizon, composed of undecomposed leaves and other organic matter. The A-horizon directly below is commonly called topsoil. As water percolates downward, minerals dissolved from the A-horizon are deposited in the B-horizon. Below is the parent material or bedrock.

Soils of the Ozark Highlands can be categorized by various means. According to their bedrock, those of the Boston Mountains were derived from sandstone and shale; those of the St. Francois Mountains and adjacent areas of southeastern Missouri were derived from several types of igneous rock. Soils of the remainder of the Ozark Plateau were derived from limestone, chert, and dolomite. The Ozark Hills of Illinois have soils derived from limestone and sandstone. In general, soils of these areas are thin and rocky, a major factor influencing their vegetation.

The principal rock formations from which soils of the Ouachita Mountains have developed include sandstone, shale, and novaculite. The latter is a very hard rock, of silica (quartz), that forms the backbone of the ridges that characterize the Ouachitas. As in the Ozark regions to the north, soils are relatively thin, making them generally unsuitable for agriculture.

Crowley's Ridge is characterized by a type of nonresidual soil known as *loess*. Such a soil, after developing elsewhere, was carried by wind to its present location during glaciation periods. Loess soils, which are often quite deep, are, however, easily eroded.

On a broader scale, most Ozarkian soils are mapped as *ultisols,* a type found also throughout most of southeastern United States and the lower elevations of the southern Appalachians. Characteristic of forested areas with a warm, relatively humid climate, ultisols are intensely weathered, as evidenced by their reddish or yellowish colors. They have a high clay content. Associated with prairies that occur throughout the Ozarks, but especially along the northern and western borders, are *mollisols*. Their dark brown color contrasts with that of ultisols; they are also deeper.

In addition to loess soils already mentioned, other nonresidual soils also occur in the Ozarks. Along streams there are *alluvial* soils, which have been transported by water. Associated with bogs or peatlands are thick, acidic *histosols,* which are rich in organic matter. Nonresidual soils are characterized by their lack of stratification or distinct horizons.

Climate and Microclimates

The Ozark Highlands, together with most of the United States to the east and south, lie within the broad climatic region designated as humid mesothermal. Such a climate is moderate in terms of both rainfall and temperature. Growing seasons are relatively long and rainfall adequate for many species of plants (although not enough for many of those of the Appalachians). Within the Ozarks, mean temperatures are highest in the south, decreasing toward the north. Rainfall decreases from east to west. As in mountains elsewhere, temperatures decrease and rainfall increases with altitude. However, altitudinal differences are less compared with those of higher mountains of the eastern and western United States.

In the Ozarks, considerable microclimatic differences—local departures from the general, prevailing climate—exist. Among the most recognized are those that exist on north- and northeast-facing slopes versus those on south- and southwest-facing ones. The latter, because they receive more direct solar radiation, tend to be considerably warmer and drier. Microclimates also result from varying soil depths, proximity to bodies of water, and other local variants from the general environment. These and other microclimates must be considered if we are to understand the conditions under which plants live, reproduce, and perform other life processes.

In this chapter we have taken an overview of the physical environment of the Ozarks. In the following ones, we will focus on its plant life and how its plants are affected by this environment.

CHAPTER 2

Forested Ecosystems of the Ozarks

An ecosystem, a local unit of nature such as a pond, forest, or prairie, consists of certain basic components. Included are biotic factors such as plants, animals, and microorganisms, and abiotic factors such as light, temperature, soil, and water. From a functional perspective, these categories of biotic components are recognized: *producers,* plants and algae that by means of photosynthesis produce the food for the entire ecosystem; *consumers,* animals that obtain their food from the producers; and *decomposers,* bacteria and fungi that break down the bodies of consumers and decomposers, making the resulting materials available for recycling. Also inherent in the ecosystem concept are numerous interspecific relationships such as predation, mutualism, and competition.

Change is a hallmark of all ecosystems. As the physical environment changes, whether slowly or more rapidly, so do the collective organisms of that ecosystem. Changes involving entire ecosystems, especially those that follow disturbances, collectively constitute *ecological succession,* or ecosystem development. Decades, or even centuries, may be required for the development of a climax or old-growth forest after a fire, logging, or the abandonment of an agricultural field.

A different kind of change involves individual plant and animal populations of an ecosystem. Long-term incremental, but permanent, changes in populations occur, often giving rise to new species. Such changes, which make up *organic evolution,* increase biodiversity, thus enhancing the stability of the ecosystem. Within various parts of the Ozarks, as elsewhere, both succession and evolution are constantly and simultaneously occurring. Because ecosystems are dynamic, these events must be considered when we consider and describe the nature of ecosystems.

Interior Highlands

The term *vegetation* refers to plants that, because of their size or numbers, dominate an ecosystem. Forests, dominated by trees, are further designated by the particular combination of tree species that are most abundant. According to the vegetation map prepared by the geographer A. W. Küchler, the entire Interior Highlands, including both the Ozark Plateau and Ouachita Mountains, lie within the oak-hickory forest region. Here occur various species of oak, hickory, other trees, and smaller associated plants, shrubs, and herbs that can tolerate considerably less rainfall than is normal in the forests east of the Mississippi River. Such a forest extends north to south along the western edge of the eastern deciduous forest, where it interfaces with the vast (or once vast) tallgrass prairie to the west. According to E. Lucy Braun and other ecologists, it is within the Interior Highlands that the oak-hickory forest reaches its highest development.

Oak trees that make up the overstory, or canopy, of these core Ozarkian oak-hickory forests include White, Black, Post Oak, Blackjack, and Shumard, among others. Common hickories are Pecan, Bitternut, and Shagbark. Flowering Dogwood and Eastern Redbud are common understory trees.

Oak-hickory forests harbor a multitude of herbs or wildflowers. Among the monocots are Fly Poison, Solomon's-seal, Turk's-cap Lily, and Large and Small Yellow Lady's-slippers. Herbaceous dicots include Rue-anemone, American Columbo, Black Cohosh, Wild Strawberry, Common Blue Violet, Woodland Sunflower, and many others.

In reality, considerable variation occurs from site to site within forests of the Interior Highlands. In some places hickories are absent; thus a mixed oak forest results. Often pines, especially Shortleaf Pine, are present, sometimes codominant with oaks or even forming local pure stands, especially in the Ouachitas. Unlike oak and hickory, which vary by species from site to site determined largely by soil moisture, pines are more likely to occur in areas of past disturbances such as fires or windstorms. On drier sites such as ridges and south-facing slopes, Post Oak–Blackjack Oak forests are common.

On mesic (moist) sites of the Ozarks, mixed mesophytic forests with many tree species sharing dominance occur. They are similar to those of the Appalachians. Rivers such as the Current, White, Osage, and smaller ones have cut valleys where their slopes provide mesophytic habitats. Trees found here that are not found in dry Ozark forests include Northern Red Oak, Basswood, and Sugar Maple (the last-mentioned especially on limestone soils). Understory trees, in addition to Flowering Dogwood and Eastern Redbud, also found in oak-hickory forests, include Umbrella and Cucumber Magnolias. Spice-bush, Wild Hydrangea, Pawpaw, and Eastern and Ozark Witch-hazel are among the shrubs.

The forest floor of Ozarkian mixed mesophytic forests is rich in wildflowers. Among monocots are the following: Jack-in-the-pulpit, White Trillium, Wake Robin, Devil's-bit, Yellow Dogtooth Violet, Putty-root, and Crane-fly

City View overlook, Ozark Plateau of northcentral Arkansas, near Calico Rock; Current River below.

Orchid. Representative herbaceous dicots include May-apple, Doll's Eyes, Green Violet, Canada and Yellow Violets, Wild Ginger, Sweet Cicely, Virginia Waterleaf, Spring Beauty, and Wild Geranium.

As we have seen, forests of the Interior Highlands vary considerably in their composition. Not only are soil moisture and other currently existing environmental factors important, but also the history of a local area must be considered. After weighing such variables, and recognizing local variations, most ecologists now designate the Interior Highlands as an oak-hickory-pine forest region. This Ozark region contains the highest plant biodiversity of the midcontinental area. In addition to the availability of the mountains for colonization and habitation for many millennia (chap. 1), there is also a great variety of microhabitats because of wide differences in topography, drainage, and geology.

Ozark Floristics

Floristically, plants of the Ozarks may be categorized according to their origins or affinities to the flora of other regions. Many are primarily *prairie* species that occur in Ozarkian prairies or other open sites. Others, characteristic of the *Coastal Plains* to the south and southeast, occur in moist, low-lying sites, especially along the southern border of the Ozarks. Many wildflowers, like some trees already mentioned, have *Appalachian* affinities. Another large group are *exotics,* plants that are not native to the Ozarks but have become naturalized, especially in disturbed soils. Many, but not all, are weeds native to Europe or Asia.

Of special interest to botanists and wildflower enthusiasts are Ozark *endemics,* plants found only in this region, where they apparently originated. Among the several dozen species in this category are the monocots Ozark Spiderwort and Ozark Wake Robin, not to mention several grass and sedge species. A few of the many endemic herbaceous dicots are Trelease's Larkspur, Missouri Bladderpod, Ovate-leaf Catchfly, and Yellow Coneflower. Woody plants include Buck Brush, Ouachita Lead Plant, Ozark Chinquapin, and Ozark Witch-hazel. Many other Ozarkian rare plants not now confined to the Ozarks, such as Alabama Snow-wreath, may well have originated there. In addition to endemic Ozarkian species, many plant varieties (subdivisions of species) are known only from the Ozarks. With the passage of time (or reevaluation by botanists), some endemic varieties will likely in the future be recognized as endemic species of the Ozarks.

It is now recognized that the Ouachita Mountains of Arkansas and adjacent Oklahoma contain an unusually large number of endemic plant species, the most of any area within the Interior Highlands province.

Ozark Hills and Crowley's Ridge

The Ozark Hills of southwestern Illinois, while considered an extension of the Ozark Plateau of Missouri, is an area of more rugged topography that supports a greater diversity of forests than does the Ozark Plateau proper. Uplands of the region are more commonly occupied by mixed oak forests, but ravines and sheltered slopes harbor mixed mesophytic tree species such as Tulip-tree, American Beech, Sugar Maple, Cucumber Tree, and others. This Ozark region includes the same floristic categories and many of the same species as outlined above for the core Interior Highlands across the river.

Crowley's Ridge has already been identified as a narrow, elevated, loess "island" within the delta region of eastern Arkansas and Missouri. It is (or was) largely covered by mixed mesophytic forests and is therefore more like the Ozark Hills across the Mississippi River than the Interior Highlands to the west. An example of this eastern affinity is the presence of Tuliptree not otherwise found west of the river. Nevertheless, many wildflowers of the ridge are typical of the core Ozark region as well as of uplands of the

Mixed mesophytic forest of LaRue–Pine Hills Ecological Area of the southern Illinois Ozarks.

Mixed mesophytic forest of Village State Park, Crowley's Ridge, eastern Arkansas.

eastern United States. Certainly, the flora of the ridge differs sharply from that of the adjacent low-lying delta.

Also of note is the occurrence of a number of bogs throughout the ridge. These wetlands, characteristic of the Northeast and Great Lakes regions, only rarely occur this far south. More information about these bogs and their plants is included in the following chapter.

CHAPTER 3

Open and Wetland Ecosystems of the Ozarks

Although the Ozark region has been recognized (chap. 2) as supporting oak-hickory-pine forests and their many variants, the actual ecosystems of the area are often quite different. Here we are considering two major types of disruption in the continuity of Ozark forests: (1) open land areas dominated primarily by shrubs or herbs, and (2) wetlands (areas covered by water for only a part of each year) and aquatic ecosystems such as streams, lakes, ponds, and bogs.

Open Ozark Ecosystems

When early explorers—and later botanists—of the Ozarks encountered open treeless areas or those with only scattered trees, they recognized them as places of special interest. Whether called barrens, glades, prairies, or other similar names, these areas often support distinctive plant communities that differ markedly from those of adjacent forests.

Savannas are openings in oak forests. They have a parklike appearance with a herbaceous ground cover and only widely spaced trees. These ecosystems were formerly prevalent in much of the Ozark Plateau of Missouri and Oklahoma (less often in Arkansas), and on Crowley's Ridge. Where soils are relatively deep, trees are (or were) principally Bur, White, and Pin Oak. On thinner soils Eastern Redcedar and Shortleaf Pine occur. Grasses and sedges are typically those of the tallgrass prairie. Forbs (wildflowers), which are also those of other open areas, are listed below.

Barrens are primarily grasslands on deep to shallow rocky soils. *Limestone glades* are areas with extensive rock outcrops and very thin soils. (Cedar or limestone glades of middle Tennessee share many of the physical features of Ozarkian glades but support a quite different flora). Ozark glades occur principally on south and southwest slopes where dolomite rocks form outcrops. Glades of southcentral Missouri, when on hilltops or hillsides, are

Bald knob overlooking Branson; Ozark Plateau of southcentral Missouri.

often called "bald knobs" by local folk. Barrens and glades may be totally open or include scattered woody plants. Trees include Eastern and Ashe's Redcedar, Eastern Redbud, Fringe Tree, American Smoke-tree, Flowering Dogwood, and various oak species. Among the common shrubs are New Jersey Tea, Carolina Buckthorn, Buck Brush, Gum Bumelia, Smooth Sumac, Winged Sumac, and Fragrant Sumac.

The herbaceous flora of savannas, barrens, and glades include many of the same species. Among the monocots are Wild Hyacinth, Spanish Bayonet, False Garlic, False Aloe, Blackberry-lily, and Slender and Nodding Ladies'-tresses. Herbaceous dicots include Thimbleweed, White Indigo, Creamy Indigo, Blue False Indigo, Pale Beardtongue, Eastern Prickly Pear, Sensitive Briar, Pencil Flower, Rose-pink, Rose Verbena, Blue Sage, Bluet, Purple Coneflower, Yellow Coneflower, Pale Coneflower, and Western Daisy.

Prairies of the Ozarks occur primarily on the Ozark Plateau, especially along its western edge, where rainfall is less frequent than farther east, but also elsewhere, including the Illinois Ozarks. Trees, if present, are more widely scattered than in savannas. In addition to many species of grasses and sedges, forbs are abundant and diverse. Among monocots are Prairie Onion and several species of blue-eyed grasses. Herbaceous dicots include Prairie Parsley, Indian Paint-brush, Goat's-rue, Downy Phlox, Butterfly-weed, Bird-foot Violet, Soapwort Gentian, and several species each of coneflowers, goldenrods, sunflowers, asters, and blazing stars. Shrubby dicots include Prairie and Pasture Rose.

Fults Prairie, Illinois Ozarks (floodplain of Mississippi River in background).

It should be apparent that these Ozark ecosystems are not rigid categories but convenient labels for natural areas that vary continuously from savannas to prairies. Many plant species, both herbaceous and woody, are shared among these intriguing habitats.

Explanations for the origin and maintenance of these ecosystems often include some combination of fire, human disturbance, or special edaphic (soil) factors. Unfortunately, fire suppression, overgrazing, and other human-related practices have greatly reduced the extent of these special places.

Ozark Wetlands and Aquatic Ecosystems

Wetlands are transitional between terrestrial and aquatic ecosystems. Once considered worthless or even sinister, swamps, marshes, and other wetlands are now recognized as valuable habitats for a multitude of plant and animal species. As they are now protected by federal legislation, their boundaries must be delineated. Wetlands are designated by the following criteria: (1) water at or near the surface for a part of each year; (2) hydric soils, azonal ones that are visually distinct from adjacent upland soils; and (3) the presence of hydrophytes, plants characteristic of wet sites. Recognition of these three features allows conservationists to identify, study, and ideally preserve these critically important ecosystems.

Ecologists often classify wetlands according to their dominant plants. *Marshes* are wetlands dominated by grasses and graminoids (rushes and sedges); *wet meadows* are similar to marshes but are covered by water less frequently. *Swamps* are dominated by trees but also harbor a diversity of shrubs and herbs.

Other wetlands are designated by physical features. *Floodplains* are overflow areas that parallel streams; their vegetation is variable. *Bogs*, rare in the Ozarks, are wetlands characterized by the dominance of sphagnum moss, which produces peat and makes the water and soil acidic.

Except for bogs, these various wetlands environments are inhabited by many of the same species of wildflowers. Monocots include Broad-leafed Arrowhead, Spider Lily, Yellow-eyed Grass, Pickerelweed, Red and Yellow Iris, Southern Blue Flag, Common Cattail, Green Wood Orchid, Lily-leaved Twayblade, and Yellow Fringed Orchid. Herbaceous dicots include Spotted Cowbane, Hooked Buttercup, Spotted and Pale Jewelweed, Wood Betony, Cardinal Flower, several species of knotweeds, Halbert-leaved and Hairy-fruited Rose Mallow, Hydrolea, Virginia Bluebells, Mistflower, and Hollow and Sweet Joe-Pye-weed. Among common wetland shrubs are Virginia Willow, Spice-bush, Indigo Bush, Strawberry Bush, Wahoo, and several species of wild grapevines. Wetlands trees vary according to region; common ones are Baldcypress and Swamp Tupelo.

Many of the wildflowers found in bogs of Crowley's Ridge are also found in wetlands of the Ozarks to the west. Among those of limited distribution

Blanchard Springs, Ozark National Forest; Ozark Plateau of northcentral Arkansas.

are the following monocots: Michigan Lily, Ragged-fringed Orchid, and Crane-fly Orchid. Rare dicots include White Turtlehead and Steeple-bush.

Aquatic ecosystems have water throughout the year. *Streams* are linear aquatic ecosystems that include a variety of niches. Near the spray zones of waterfalls are often a variety of nonflowering plants—mosses, liverworts, and ferns as well as moisture-requiring wildflowers. Wildflowers are not as numerous in the deeper waters of streams, lakes, and ponds as they are in wetlands. In shallow moving water, Water-willow is often found. American Lotus, Fragrant Water-lily, and Spatterdock are among plants seen floating on the open water of ponds and lakes. Along the margins and floodplains of these aquatic ecosystems, wildflowers are much the same as in wetlands.

Conservation of Ozark Ecosystems

European settlers of North America considered forests enemies to be conquered and subdued. Consequently, they cleared the land and used forest resources as if they were limitless. By the time people recognized that few untouched old-growth forests and other pristine ecosystems remained, it was almost too late.

In the Ozarks as well as elsewhere, extinction of species is an ever-present concern. As loss of habitat is the greatest threat to rare and endangered plants, it is imperative that we preserve representative ecosystems of each

Baker Prairie, a natural area near Harrison; Ozark Plateau of northcentral Arkansas.

type—terrestrial and aquatic. In so doing, we will preserve the biodiversity of the Ozarks, including plant and animal species as well as the ecosystems themselves. Fortunately, human population pressure in most Ozark regions is less intense than in many other parts of the country, making it feasible to implement such critical measures before it is too late.

At a more personal and immediate level, we must recognize that some of the threats to wildflowers are due to the activities of wildflower enthusiasts themselves. Because the digging of rare plants can reduce their numbers, plants for wildflower gardens should be obtained from nurseries that grow them from seed. Students unnecessarily take some plants and press them for school projects. Although scientific studies often require that specimens be taken and deposited in a herbarium, in most cases color photographs are satisfactory for documentation.

CHAPTER 4

Identifying Ozark Wildflowers

Identification of a plant or animal is the initial step in the process of learning about that organism. Whether one learns the common or scientific name, a wealth of knowledge about the organism then becomes available.

Naming Plants

Common names of plants are often fanciful or descriptive. Examples of common names (with scientific names in parentheses) include Spring Beauty (*Claytonia virginica* L.), Ozark Spiderwort (*Tradescantia ozarkana* Anderson & Anderson), and Yellow Dog-tooth Violet (*Erythronium rostratum* Wolf). The disadvantage of using common names is that many species have more than one common name and the same name is often applied to more than one species.

To avoid confusion, botanists and zoologists use Latin or Latinized scientific names for each species. Under this system, established by the Swedish botanist Carolus Linnaeus in the eighteenth century, each species is assigned a binomial scientific name. The *genus* name (a noun) is followed by the *specific epithet* (an adjective); both are italicized. The name, names, or initials that follow indicate the person or persons who first assigned the scientific name. As Linnaeus assigned the names of many common plants of North America, his initial (L.) is often used. When, as occasionally happens, a name must be changed, the name of the earlier author appears in parentheses. For example, the scientific name of Western Tickseed appears as *Bidens aristosa* (Michx.) Britt. This species was first assigned to one genus by the botanist Andre Michaux; it was later placed in the genus *Bidens* by Nathaniel L. Britton. When a trinomial is used as a scientific name, the third italicized word indicates the *variety* or *subspecies* within the species to which the plant belongs (example: *Viola canadensis* var. *canadensis*).

What Is a Wildflower?

Wildflowers are flowering plants (botanists use the word *angiosperm*) that grow wild within a particular area. Most are native plants, but others are exotic, often weeds of Eurasian or (less often) tropical American origin.

Whereas some consider only herbaceous (nonwoody) plants to be wildflowers, this book includes also flowering shrubs, trees, and vines, especially those with conspicuous flowers. Grasses and their relatives, sedges and rushes, which have tiny inconspicuous flowers that are difficult to identify, are not featured here.

Leaves and Flowers

All parts of a plant are sometimes used for identification. Leaves and flowers, because they are most often of importance, are emphasized here.

Leaves, the principal light-collecting organs of a plant, vary greatly in size, shape, and arrangement, reflecting the evolutionary history of the plant as well as the immediate environment. Some common leaf variations are shown in fig. 1.

FIGURE 1. LEAF STRUCTURES AND VARIATION

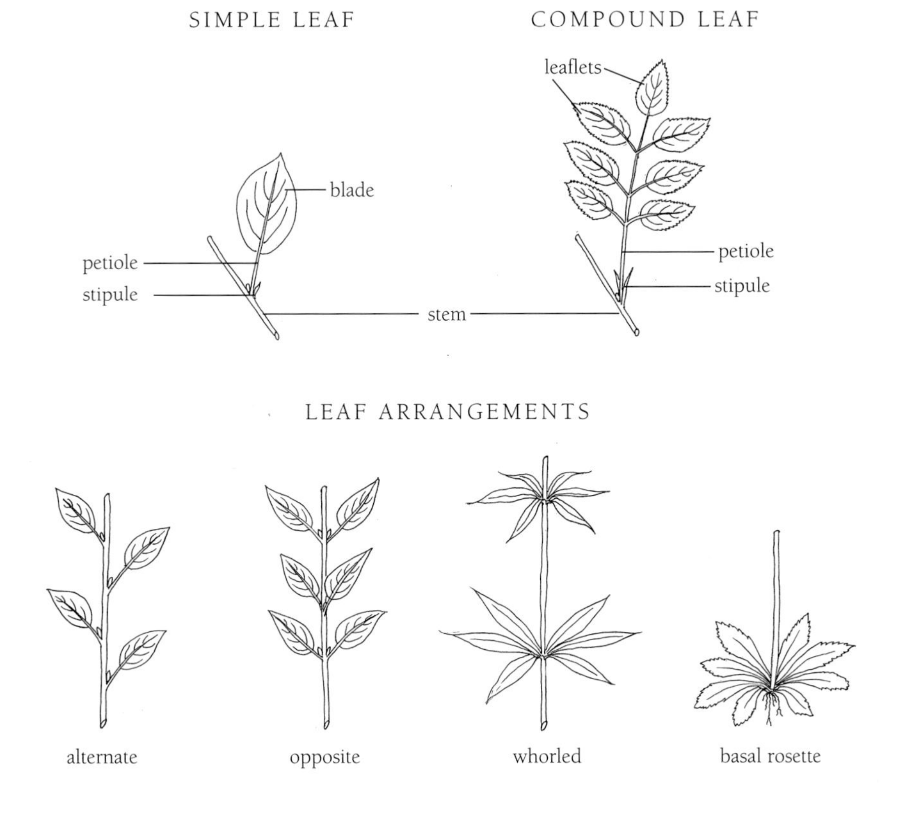

LEAF SHAPES

linear

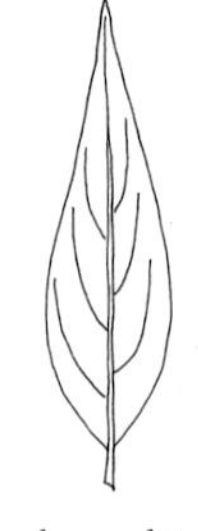
lanceolate

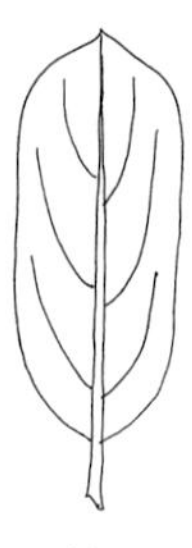
oblong

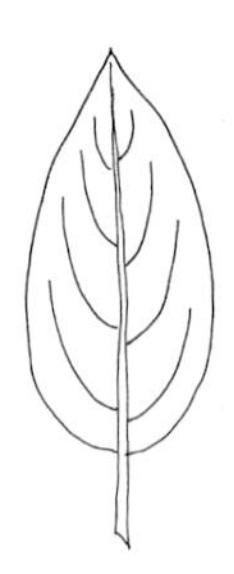
ovate

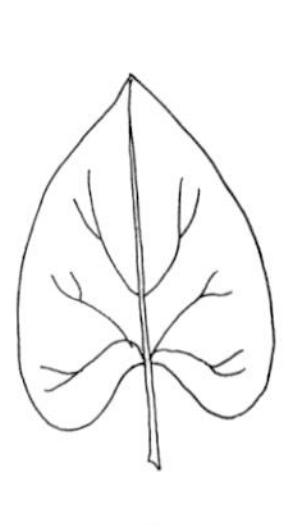
cordate

BLADE MARGINS

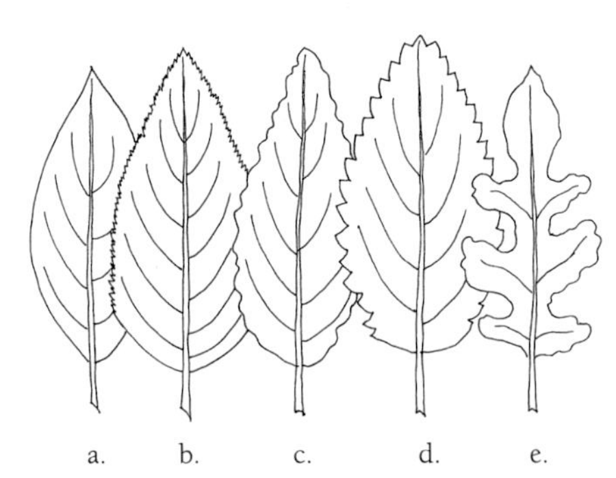

a. entire
b. serrate
c. undulate
d. dentate
e. lobed

VENATION

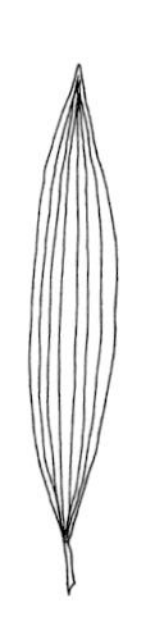
parallel-veined

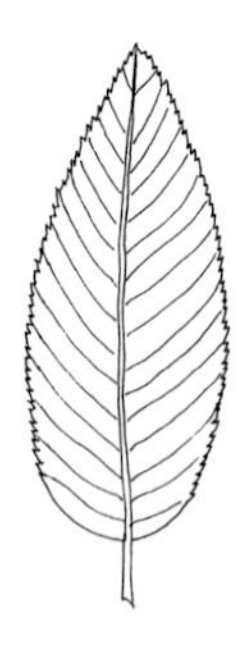
pinnately net-veined

palmately net-veined

Flowers contain sexual parts, allowing them to serve primarily a reproductive function. As floral structures reflect kinship between species, they are especially useful for identification and assignment of species to taxonomic categories such as genera, families, and orders.

A typical, or representative, flower (fig. 2) includes four whorls—concentric rings of parts. From the outside inward:

Sepals (collectively, calyx) are usually green but often petal-like (white or some color other than green).

Petals (collectively, corolla) are often the showiest parts of the flower; they may be united to form tubular or funnel-like structures.

Each *stamen* is composed of a knoblike anther, which produces pollen, supported by a slender filament.

The *pistil* (one or more) consists of (from bottom up) an ovary, a style, and a stigma.

Flowers vary greatly. In some, stamens are present, but pistils are absent; such a flower is said to be staminate. A flower that has only pistils is called a pistillate flower. Flowers with both stamens and pistils are bisexual. Monoecious plants have both staminate and pistillate flowers on the same individual, whereas those with the two kinds of flowers on separate individuals are dioecious.

In many plants, flowers occur in arrangements known as inflorescences. Some of the more common ones are shown diagrammatically in fig. 2. In each case a common stalk, the peduncle, supports a cluster of flowers. In some instances, a single type of inflorescence is associated with a particular plant family. For example, each common daisylike "flower" of the aster family (Asteraceae) is actually a head consisting of numerous tiny flowers (florets). The flowers of many plants of the pea family (Fabaceae) also have florets in heads (for example, clovers).

Monocots versus Dicots

All flowering plants belong to one of two grand groups of Angiosperms: monocots (Monocotyledones) or dicots (Dicotyledones). These names are based on the number of cotyledons (one or two) possessed by the seed embryo. As seeds are often small and their cotyledons are not easily seen, it is more practical to rely instead on leaf or flower features or both to determine to which of these groups a given plant belongs.

The leaves of monocots are typically parallel-veined, whereas those of dicots are either pinnately or palmately net-veined (see fig. 1). The flowers of monocots are 3-merous; that is, each or most whorls contain 3 (or twice 3) parts (see fig. 1). An example is a lily with 3 sepals, 3 petals, 6 stamens, and 1 pistil. The basic number for flowers of dicots is something other than 3; usually they are either 4- or 5-merous.

FIGURE 2. FLOWER STRUCTURES AND ARRANGEMENTS

TYPICAL FLOWER

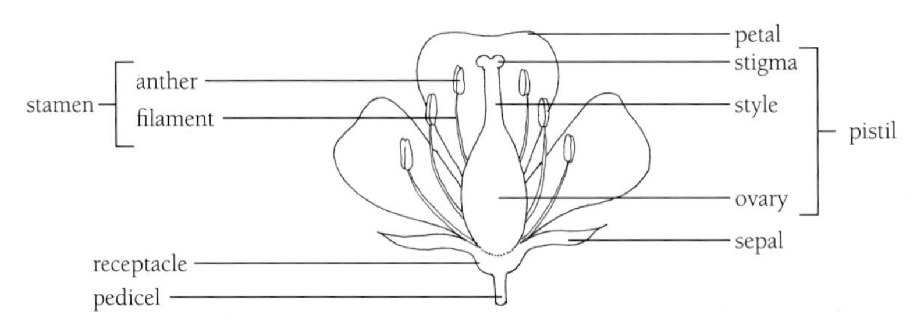

INFLORESCENCES

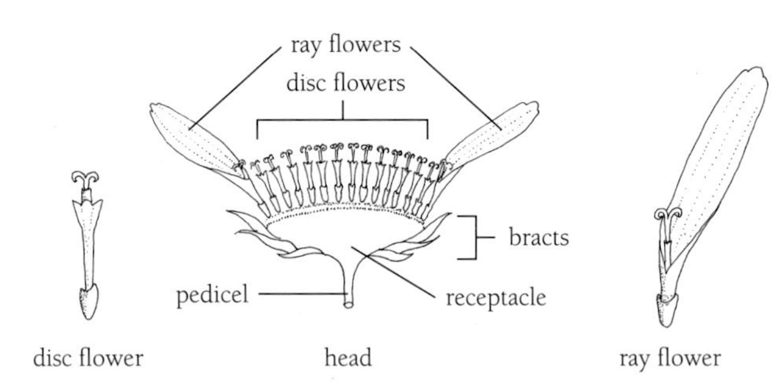

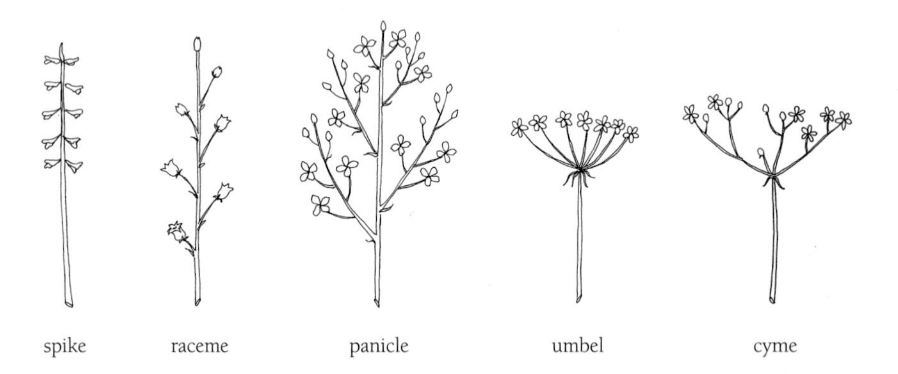

Part 2

Ozark Wildflowers Illustrated

Using the Color Plates

Identifying a wildflower allows you to call it by name (common, scientific, or both), thereby opening up a wealth of information about that particular plant.

The initial step is to recognize to which of the sixteen categories your unknown plant belongs (see table of contents and "Using This Book"). The categories are based primarily on flower (or fruit) color and secondarily on the grand groups of plants known as monocots or dicots. (Within each of the categories, plants are arranged by families.)

In addition to the more than 250 species pictured and described in some detail, many related species are compared. Thus more than 600 species of Ozark plants can be identified using this book.

"Flower" refers to the showiest part of a plant, usually sepals or petals of the flower itself, but sometimes colorful bracts (modified leaves) associated with the flowers. In the case of flower "heads," such as those of daisies, the color of the petals of the ray florets that surround the head are to be used. For bicolored flowers, the darker or more prominent color is used in the key.

Defining Terms

"Fruit" is defined as a ripened ovary with enclosed seeds; examples are berries and capsules.

"Woody plants" include trees, shrubs, and vines that develop hard, woody stems.

"Herbaceous plants" (herbs) are those with stems that lack woody tissues.

"Monocots" are plants with parallel-veined leaves and floral parts in 3s (example: 3 sepals, 3 petals, 6 stamens, 1 pistil).

"Dicots" are plants with net-veined leaves and floral parts in 4s or 5s (example: 4 sepals, 4 petals, 8 stamens, 1 pistil).

A rather detailed description of flowers, leaves, and terms used to describe them is given in chapter 4, "Identifying Ozark Wildflowers." Other terms are defined in appendix 1, "Glossary."

White

Ozark Spiderwort *Tradescantia ozarkana* Anderson & Woodson
Spiderwort Family Commelinaceae

Spiderworts are perennials with terminal clusters of flowers, each about 1 in. across. Petals colors vary from blue to purple, rose, or, less often, white.

Ozark Spiderwort grows to 2 ft. tall. Its stems and leaves are silvery green with a whitish coating that is easily rubbed off. Flower color varies from the nearly white seen here to light blue or lavender. It is an uncommon plant of steep, shaded limestone bluffs of the Ozark Plateau of MO and AR; also the Ouachita Mountains of AR; Apr., May.

Woodland Spiderwort, *T. ernestiana* Anderson & Woodson, has smooth stems and leaves lacking the whitish coating; its flower petals are deeper shades of rose, blue, or purple. It sometimes hybridizes with Ozark Spiderwort; Apr., May.

Broad-leaved Arrowhead *Sagittaria latifolia* Willd.
Water Plantain Family Alismataceae

Arrowheads are common aquatic plants; each of their flowers has 3 white, rounded petals. Leaf shape varies with species; in this one they are 8–10 in. long and very broadly arrow-shaped. Also called Common Arrowhead, the plant is found throughout the Ozark region; Jun.–Oct.

Grass-leaved Arrowhead, *S. graminea* Michx., with leaves consistent with its name, is also found throughout our area but is less common; May–Aug.

White Trillium *Trillium flexipes* Raf. [*T. gleasoni* Fern.]
Lily Family Liliaceae

The lily family is a large and widespread group of plants, mainly of temperate regions. A typical flower includes 3 sepals, 3 petals, 6 stamens, and a long pistil with a 3-lobed stigma at its end. Nearly all are perennials that overwinter as bulbs.

This, the largest of Ozark trilliums, reaches 2 ft. in height and has 3 broad leaves, each 5 in. across. The single, usually nodding, flower is on a long stalk (to 4 in.) above the leaves. The fruits turn a bright rose color when mature. Also called White Wake Robin, it is an uncommon plant of moist, rich woods, often found in ravines or along streams. In our area it occurs along the e. edge of the Ozark Plateau of MO and in Stone Co. of n. AR; also in IL Ozarks; Apr., May.

Ozark Wake Robin, *T. pusillum* var. *ozarkanum* (Palmer & Steyerm.) Steyerm., is found sporadically in dry oak woodlands of the w. Ozark Plateau (MO, AR) and also w. Ouachitas (AR, OK). It is a smaller plant with narrow leaves and erect flowers that have white petals with crinkled margins. Petals change to pink as they age; Apr., May.

Ozark Spiderwort
Tradescantia ozarkana

Broad-leaved Arrowhead
Sagittaria latifolia

White Trillium
Trillium flexipes

Fly Poison *Amianthium muscaetoxicum* (Walt.) Gray
Lily Family Liliaceae

The upright or bending stalks (to 4 ft.) of Fly Poison bear small white flowers that turn green to purple as they age. Basal leaves are linear and up to 2 ft. long. It is found in sandy or acidic soils in woods of the Ozark Plateau and Ouachita Mountains of AR and OK; May–Jul.

All parts of the plant, but especially the bulbs, contain poisonous alkaloids. Colonists mixed a paste made from the bulbs with sugar to kill flies.

Devil's-bit *Chamaelirium luteum* (L.) Gray
Lily Family Liliaceae

Also called Fairy-wand, this uncommon perennial has curved spikes (4–5 in. long) with tiny greenish or white flowers that turn yellowish with age. Male and female flowers are on separate plants. Basal leaves form a rosette; several smaller leaves are arranged along the flowering stalk. It occurs in rich woods of the Ouachita Mountains of AR; Mar.–May.

Colic-root *Aletris farinosa* L.
Lily Family Lilaceae

A tall (2–3 ft.) stalk bears the small urn-shaped flowers, each swollen at the base. Long, sharp-pointed, lanceolate leaves form a basal rosette. The perennial is found in sandy or peaty soils of the se. Ouachita Mountains of AR; Apr.–Jun.

A root decoction has been used to treat indigestion, colic, and other ailments. It contains diosgenin, known to be anti-inflammatory.

Yellow Colic-root, *A. aurea* Walt., is a similar but yellow-flowered plant of w. AR and e. OK; Apr.–Jul.

Spanish Bayonet *Yucca smalliana* Fern. [*Y. filamentosa*]
Lily Family Liliaceae

The several *Yucca* species of N. Amer. bear various names such as Beargrass, Spanish Bayonet, and Soapweed. All have white or whitish flowers arranged on a tall, thick stalk. Each flower includes 6 tepals: 3 sepals and 3 petals. At the base of the flower stalk is a rosette of leathery, sharply pointed leaves.

Spanish Bayonet has a tall (6–9 ft.) panicle of greenish white flowers. Typically, there are loose threads attached to the leaves. Native to the Coastal Plain of e. U.S., it is seen only sporadically as an escape in open, sunny places of our area; May–Aug.

Arkansas Yucca, *Y. glauca* var. *mollis*, has similar flowers and leaves but is distinguished by its arrangement of flowers without prominent side branches; May–Jun.

Fibers from the leaves of *Yucca* plants have been used for cord.

Fly Poison
Amianthium muscaetoxicum

Devil's-bit
Chamaelirium luteum

Colic-root
Aletris farinosa

Spanish Bayonet
Yucca smalliana

Solomon's-seal *Polygonatum biflorum* (Walt.) Ell.
Lily Family Liliaceae

This 3- to 4-ft.-tall perennial with an arching stem has smooth alternate leaves and 8–12 white to greenish flowers arranged in pairs (*biflorum*) along its nodes. Berries that follow are brownish. It occurs in a wide range of dry to moist woods throughout most of our area; May, Jun.; fruits, Jun.–Oct.

Some botanists consider plants of Solomon's-seal with wider leaves to be a separate species, *P. canaliculatum* (Muhl.) Pursh; May, Jun. Plants of the two types often hybridize, resulting in plants with intermediate traits.

A root tea was widely used by both Native Americans and white settlers for a variety of digestive and joint disorders. Modern research, however, does not support these uses. The young leafy shoots can be eaten as a potherb.

Solomon's-plume *Maianthemum racemosum* (L.) Link [*Smilacina racemosa* (L.) Desf.]
Lily Family Lilaceae

In contrast to Solomon's-seal (above), this species has numerous tiny, white, starlike flowers in a terminal panicle. Fruits are red berries flecked with purple. The stem is somewhat zigzag. Also called False Solomon's-seal, it is found in much of the same habitats and range as Solomon's-seal; May, Jun.; fruits, Jun.–Oct.

Native Americans used a leaf tea as a contraceptive and for coughs; root tea, as a stomach tonic and purgative. The young shoots can be cooked and eaten like asparagus or added to salads.

Starry False Solomon's-seal, *M. stellata* (L.) Desf., is similar but has black berries and larger flowers in smaller clusters; May, Jun.; fruits, Jun.–Oct.

Star-of-Bethlehem *Ornithogalum umbellatum* L.
Lily Family Liliaceae

This plant, 4–12 in. tall, has flowers about 1 in. across. The grasslike leaves have whitish midribs (main veins). Introduced from Europe, it is found in lawns, along roadsides, and in other disturbed places throughout our area; Apr.–Jun.

Glycosides, toxic compounds, are located in all parts of the plant, especially the bulbs, and may cause serious illness (even death) if ingested by humans (most cases involve children). Cattle have also been poisoned from eating the plant.

Solomon's-seal
Polygonatum biflorum

Solomon's-plume
Maianthemum racemosum

Star-of-Bethlehem
Ornithogalum umbellatum

False Garlic *Nothoscordum bivalve* (L.) Britt. [*Allium bivalve* (L.) Kuntze]

Lily Family Liliaceae

The white or cream flowers, each ½ in. wide, are arranged in clusters of 3–10 at the top of a 1-ft.-tall stem. Leaves, which are separate from the flowering stalks, are linear. A common weedy plant, it occurs generally throughout our area, especially on limestone glades and in sandy soils along streams and wet prairies; Mar.–May (often again in autumn).

Spider Lily *Hymenocallis caroliniana* (L.) Herb. [*H. occidentalis* (Le Conte) Kunth]

Amaryllis Family Amaryllidaceae

This striking perennial, which grows from a large bulb, has linear basal leaves. The flower stalk, which may be 3–4 ft. tall, bears several (3–7) flowers in a large umbel. From the long slender corolla tube extend 6 narrow tepals, which are connected by the round membranous crown; attached to it are the 6 long stamens. Spider Lily is principally a plant of low wetlands e. and s. of our area, but also occurs in wet woodlands along the e. edge of the Ozark Plateau, on Crowley's Ridge, and in the IL Ozarks; Jul., Aug.

In central AR, the range of *H. liriosme* (Raf.) Shinners, also called Spider Lily, overlaps with that of *H. caroliniana,* and the two species hybridize. They look very similar, but *H. liriosme* blooms somewhat earlier; Jun., Jul.

Slender Ladies'-tresses *Spiranthes lacera* (Raf.) Raf. [*S. gracilis* (Bigel.) Beck]

Nodding Ladies'-tresses *Spiranthes cernua* (L.) Richard

Orchid Family Orchidaceae

The approximately 8 Ozark species of *Spiranthes* are recognized by the spiral arrangement of their tiny white or greenish flowers; they flower principally in the late summer or fall. Basal leaves, which appear in spring, usually wither before flowering time. They generally occupy limestone glades, open woods, and thickets, including a wide range of wet to dry conditions. Various species have been used as diuretics and aphrodisiacs.

In Slender Ladies'-tresses (left), the slender flower stalk is 20–30 in. tall, and the flowers are arranged in a single spiral. It occurs throughout the Ozarks, except for Crowley's Ridge; Aug.–Oct.

The white flowers of Nodding Ladies'-tresses (right) are more compact, and each flower nods (bends downward) about 45° from the horizontal. The most common species of the genus, it is seen, often in large colonies, throughout our area; Aug.–Nov.

Shining Ladies' Tresses, *S. lucida* (Small) Ames, differs from other Ozarkian *Spiranthes* species by the pale orange lip of each flower and its shiny leaves still present at flowering; May, Jun.

False Garlic
Nothoscordum bivalve

Spider Lily
Hymenocallis caroliniana

Slender Ladies'-tresses
Spiranthes lacera
Nodding Ladies'-tresses
Spiranthes cernua

Fragrant Water-lily *Nymphaea odorata* Ait.
Water-lily Family Nymphaeaceae

Water-lilies are showy plants of shallow water; their flowers and leaves are attached by stems to roots that anchor them in the soil. In this species leaves are large, round, and notched at the base. Flowers are 5 in. across, white (or red in form *rubra*), and fragrant. It is restricted to ponds, lakes, and slow-moving streams throughout our area; Jul.–Aug.

Native Americans poulticed the roots to reduce inflammation; roots were also used to make tea for coughs and tuberculosis.

Other water-lilies are featured in the yellow section.

Sandwort *Arenaria patula* Michx.
Pink Family Caryophyllaceae

Sandworts are well named; they are small plants typically found in sandy soil, where they form extensive mats. This small annual, 3–6 in. tall, has opposite needlelike leaves. Note the 5 notched petals, each with tiny green stripes and about ½ in. long. It occurs in glades and rocky prairies throughout the Ozarks except for the IL Ozarks and Crowley's Ridge of AR; Apr.–Jun.

Thyme-leaved Sandwort, *A. serpyllifolia* Michx., a European native, is also widespread and common. It has tiny, narrowly lanceolate leaves, also arranged in pairs; Apr.–Aug.

May-apple *Podophyllum peltatum* L.
Barberry Family Berberidaceae

May-apple has shiny, paired, umbrella-like leaves, 8–10 in. across. Their solitary flowers are in axils of the leaves. Each cuplike flower, 1–2 in. wide, has 6–9 waxy petals and twice as many stamens. Also known as Mandrake (but not to be confused with a European plant of the same name), it occupies moist woods throughout the Ozarks; Mar.–May; fruits, Jul., Aug.

The pencil-thin rhizomes have a strong purgative effect; they also contain anticancer substances and were used for this purpose by Native Americans, as they also are in modern medicine (small-cell lung and testicular cancers). Ripe fruits can be eaten or used to make a flavorful jelly, but unripe fruits and other parts are poisonous.

Fragrant Water-lily
Nymphaea odorata

Sandwort
Arenaria patula

May-apple
Podophyllum peltatum

Rue-anemone *Thalictrum thalictroides* Eames & Boivin
[*Anemonella thalictroides*]
Buttercup Family Ranunculaceae

The buttercup or crowfoot family is a large and diverse one, well represented in our area as in most temperate regions of the world. "Crowfoot" refers to the leaves, which are typically incised like a rooster's foot. Clustered in the center of each flower are numerous pistils surrounded by stamens. In many species petals are absent, the sepals resembling petals. In some, flowers are spurred.

This small (6 in. tall), dainty perennial has flowers ½–1 in. across), each with 5–10 petal-like sepals varying in color from white to light pink. Petals are absent. Note the 3-lobed, whorled, compound leaves. It occurs in dry, open woods throughout the Ozarks; Mar.–Jun.

False Rue-anemone, *Isopyrum biternatum* (Raf.) T. & G., sometimes mistaken for Rue-anemone, also occurs throughout our area but is less common and inhabits woods that are more moist. It is best distinguished by its more deeply lobed leaves and flowers with fewer sepals; Mar.–Jun.

Thimbleweed *Anemone virginiana* L.
Buttercup Family Ranunculaceae

Like Rue-anemone (above), *Anemone* species also lack petals; their white sepals (usually 5) resemble petals. Numerous pistils and stamens are spirally arranged on a central cone.

Thimbleweed is a rather tall (2–3 ft.) plant with prominently veined, deeply incised, stalked leaves. The flowers, with white or greenish petals, are not showy. The common name is due to the shape of its thimblelike fruits. It is seen in open woods throughout our area; Apr.–Aug.

Prairie Anemone, *A. caroliniana* Walt., is a shorter plant (1 ft.) with a single, larger flower (1½ in. across); stamens are orange. It occupies acidic soils of glades and prairies, primarily the w. parts of the Ozark Plateau and Ouachita Mountains of AR and OK; Mar.–May. The similar but less common Anemone, *A. berlandieri* Pritzel, has greenish or yellowish stamens; Mar.–Apr. Wood Anemone, *A. quinquefolia* L., is a low-growing (6 in. tall) plant; its leaves are typically divided into 5 leaflets; Apr.–Jun.

Black Cohosh *Cimicifuga racemosa* L.
Buttercup Family Ranunculaceae

Also called Bugbane, this tall (3–8 ft.), often branching plant bears elongated racemes of small white flowers. Leaves are three-parted with sharp teeth. It occurs in rocky woods and below bluffs of the Ozark Plateau; May–Aug.

A tincture made from the plant has been traditionally used to treat rheumatism, snakebites, and other disorders.

Rue-anemone
Thalictrum thalictroides

Thimbleweed
Anemone virginiana

Black Cohosh
Cimicifuga racemosa

Doll's-eyes *Actaea alba* (L.) Mill. [*A. pachypoda* Ell.]
Buttercup Family Ranunculaceae

Also called White Baneberry, this 1- to 2-ft.-tall perennial has racemes of white flowers followed by clusters of shiny white berries. Leaves are divided into several, often 7, leaflets with dentate margins. Look for it in rich woods and in ravines, especially on n.-facing slopes throughout the Ozarks; May, Jun.; fruits, Jul.–Oct.

The common name refers to the previous use of the shiny white berries in homemade dolls.

Dutchman's-breeches *Dicentra cucullaria* (L.) Bernh.
Fumitory Family Fumariaceae

Dicentra species have highly dissected fernlike leaves and distinctively shaped flowers. In Dutchman's-breeches, each ½-in.-long inverted flower has two inflated spurs that give it a "pantaloons" appearance, thus its common name. Its usual habitat is moist, rich woods near bluffs or on slopes, where it often occurs in large masses. It is seen throughout most of the Ozarks; Mar.–May.

Squirrel-corn, *D. canadensis* (L.) Bernh., was formerly known in Taney Co. (MO) of the Ozark Plateau, but was apparently eliminated from that site by the construction of Bull Shoals Dam on the White River. It has almost identical foliage, but its white flowers are fragrant and more heart-shaped, and lack prominent spurs; Apr., May.

Five-parted Toothwort *Cardamine concatenata* (Michx.) O. Schwarz [*Dentaria laciniata* Muhl.]
Mustard Family Brassicaceae

This is perhaps the most common of our toothworts, low, early-flowering plants with racemes of 4-petaled white to pink petals. This variable species, about 1 ft. tall, is characterized by deeply incised, whorled leaves. It occurs throughout our area, most often in rich woods; Feb.–May.

Both "toothwort" and "*Dentaria*" refer to the traditional practice of chewing the pungent roots as a toothache remedy. They also add a spicy taste to salads.

One-flowered Leavenworthia *Leavenworthia uniflora* (Michx.) Britt.
Mustard Family Brassicaceae

This plant resembles the toothwort (left) but is smaller (4–6 in. tall) and has fewer and much smaller flowers. Note the basal rosettes of dark green, deeply incised leaves. Above are several siliques (seedpods) forming from flowers that bloomed a few days earlier. It occupies glades, barrens, and bald knobs of the Ozark Plateau of MO, AR, and e. OK; Mar., Apr.

The plant is a winter annual; its seeds germinate in the fall, after which the small seedlings overwinter.

Doll's-eyes
Actaea alba

Dutchman's-breeches
Dicentra cucullaria

Five-parted Toothwort
Cardamine concatenata

One-flowered Leavenworthia
Leavenworthia uniflora

Bloodroot *Sanguinaria canadensis* L.
Poppy Family Papaveraceae

The 6- to 12-in. perennial has a lobed leaf that surrounds the flower stalk. It is uncommon but occurs in all regions of the Ozarks (less often on Crowley's Ridge than elsewhere). Its habitats include rich wooded slopes and ravines as well as limestone glades; Mar., Apr.

The thick horizontal rhizome contains bright orange red juice widely used by Native Americans to decorate their skin and to treat a variety of ailments. Like a number of other plants used for dye purposes, it was called "Puccoon."

Early Saxifrage *Saxifraga virginiensis* Michx.
Saxifrage Family Saxifragaceae

Panicles of small (¼ in. across) white flowers are borne on ft.-long stalks. Like most other members of this family, each flower has 5 sepals, 5 petals, and 10 stamens. Toothed ovate leaves, dark green above and purplish underneath, form a rosette at the base of the flower stalk. It is a plant of moist, shaded ledges and bluffs (invariably sandstone or granite). It occurs throughout most of the Ozark Plateau of MO, less commonly in AR and the IL Ozarks (apparently absent from OK); Feb.–Jun.

Two other species, both of the Ozark Plateau, resemble Early Saxifrage; plants of both have their flowers clustered near the top of the flower stalk. Texas Saxifrage, *S. texana* Buckl., has shorter petals, only barely longer than the sepals; Apr., May. Palmer's Saxifrage, *S. palmeri* Bush, is distinguished by its almost entire leaf margins; Mar.–May.

Swamp Saxifrage, *S. pensylvanica* L., is found in the ne. Ozark Plateau of MO and across the river in the IL Ozarks. It is taller than Early Saxifrage and has much larger leaves with almost entire (untoothed) margins; Apr.–Jun. Some botanists recognize as a separate species plants with very hairy leaves but otherwise similar to Swamp Saxifrage: Forbe's Saxifrage, *S. forbesii* Vasey.

Wild Strawberry *Fragaria virginiana* Duchesne
Rose Family Rosaceae

Like the cultivated strawberry, this native plant is a sprawling herb that spreads by runners. Note the large teeth on the rounded ends of the 3 dark green leaflets. Flower petals are about ¼ in. long. It is found along roads and in open woodlands throughout the Ozarks except for Crowley's Ridge; Apr., May; fruits, May.

Wild strawberries are smaller than those of cultivated varieties but are much tastier.

Indian Strawberry, *Duchesnea indica* (Andr.) Focke, an Asian species, often occurs in waste places. It has leaves like those of *Fragaria,* but its flowers are yellow and its smaller red fruits are practically inedible; Apr.–Jun.; fruits, May–Sep.

Bloodroot
Sanguinaria canadensis

Early Saxifrage
Saxifraga virginiensis

Wild Strawberry
Fragaria virginiana

Bowman's-root *Porteranthus trifoliata* (L.) Britt.
[*Gillenia trifoliata* (L.) Moench]
Rose Family Rosaceae

Bowman's-root is a smooth, 2- to 3-ft.-tall herbaceous perennial. It has leaves divided into 3 sharply pointed leaflets and flowers with 5 narrow petals. In the Ozarks it occurs sporadically only on the Ozark Plateau of MO and AR; May, Jun.

American Ipecac, *P. stipulatus* (Muhl. ex Willd.) Britt., is very similar, but it has 2 large stipules at the base of each leaf, giving the effect of 5 leaflets. Much more common, it occurs throughout most of our region, where it is at times the most conspicuous plant in flower; May, Jun.

Because of their widespread medicinal uses, both *Porteranthus* species are called Indian Physic. A leaf tea has been used to treat colds, asthma, and indigestion; a poultice, to treat insect stings and rheumatism.

Goat's-beard *Aruncus dioicus* (Walt.) Fern. var. *pubescens* (Rydb.) Fern.
Rose Family Rosaceae

This large (to 6 ft. in height), shrublike plant includes large clusters of tiny white or cream flowers held above its large compound leaves. Staminate (male) and pistillate (female) flowers are on separate plants, thus the species name *dioicus*. The variety name indicates that it is a more hairy plant than those e. of the Mississippi River. It occupies moist sites along bluffs, stream banks, and in woodlands throughout the Ozarks; May, Jun.

Native Americans poulticed roots of Goat's-beard on bee stings. A root tea was used internally for the pain of childbirth and externally to bathe swollen feet.

White Wild Indigo *Baptisia alba* (L.) Vent. [*B. leucantha*]
Pea Family Fabaceae

Baptisia species, Wild Indigo, are smooth shrublike plants with compound leaves. Pealike flowers are arranged into elongated racemes, which may be erect or sprawling.

White Wild Indigo is a tall (to 5 ft.) plant of prairies, glades, and other open places. Our only white *Baptisia* species, it occurs throughout (rare in OK); May–Jul.

Long-bracted Wild Indigo, *B. bracteata* Muhl. ex Elli. var. *glabrescens* (Larisey) Isely, is a common roadside plant, especially in acidic soils, throughout the Ozarks except for Crowley's Ridge, where it is less common. It is a spreading plant with cream flowers borne in horizontal or arching racemes; Apr.–Jun. Yellow Wild Indigo, *B. sphaerocarpa* Nutt., has bright yellow flowers in erect racemes. It occurs in the Ouachita Mountains of AR and OK and along the e. edge of the Ozark Plateau of AR; Apr.–Jun.

Bowman's-root
Porteranthus trifoliata

Goat's-beard
Aruncus dioicus

White Wild Indigo
Baptisia alba

Canada Violet *Viola canadensis* L. var. *canadensis*
Violet Family Violaceae

Canada Violet is a "stemmed" violet (leaves and flowers on same stalk) with cordate leaves and flowers on short stalks. The white petals are yellow at their bases. Its stems are purplish, and there are purplish tinges on the back of the petals (seen here in bud at right); entire petals generally turn blue or purple as they age. Canada Violet is seen in our area only in the Ozark Plateau of AR, where it inhabits rich, moist woods; Apr.–Jul.

The occurrence of Canada Violet, primarily a northern species characteristic of the cool climate of Canada and the Appalachian Mountains, in AR but not in MO is puzzling. A tea made from the roots of various violet species has been used by Native Americans for pain in the pelvic region. It was also poulticed for boils and skin wounds.

Other violets are featured in the yellow and blue/purple sections.

Pale Violet *Viola striata* Ait.
Violet Family Violaceae

Pale Violet, also a stemmed violet, can be distinguished from the much less common Canada Violet (above) by its petals, which lack the yellow and purplish marking; stems, which are smooth (not hairy); longer flower stalks; and large, deeply cut stipules. Also called Cream Violet, it occurs in all portions of the Ozarks except for OK and Crowley's Ridge; Apr.–Jun.

The less common Lance-leaved Violet, *V. lanceolata* L., is a white-flowered, stemless (flowers, leaves on separate stalks) violet. Found in swamps and other wet places, it is distinguished primarily by its lanceolate leaves, which taper toward the base; Apr.–Jun.

Indian Pipe *Monotropa uniflora* L.
Wintergreen Family Pyrolaceae

Monotropa species are saprophytic; lacking chlorophyll, their roots obtain nutrients from decomposing organic matter in the soil, which is made possible by mycorrhizal (fungus-root) relationships.

Indian Pipe is a small plant (3–8 in. tall) with a translucent stem covered with scalelike leaves. The single nodding flower turns from white to pink to black as it ages. Also called Ice Plant and Ghost Flower, it is found in ravines, upland woods, usually in thick deciduous mulch. Though not common, it occurs in all portions of the Ozarks (except for OK); Aug.–Oct.

Both Native Americans and Western physicians have used Indian Pipe for a variety of ailments. Among them are as a sedative, as an analgesic, and for the treatment of other nervous disorders. Water extracts are now known to be antibacterial.

Pinesap, *M. hypopithys* L., is a similar, less widespread plant; it has yellowish to reddish stems and flowers in racemes; Jun.–Oct.

Canada Violet
Viola canadensis

Pale Violet
Viola striata

Indian Pipe
Monotropa uniflora

Spikenard *Aralia racemosa* L.
Ginseng Family Araliaceae

Aralia species are small, generally herbaceous woodland plants with large aromatic roots, compound leaves, and tiny white flowers in clusters.

Spikenard, which grows to 4 ft., has leaves divided into 5–21 dentate leaflets and flowers in small umbels arranged into racemes. It is found in moist woods, often on n.-facing slopes of bluffs throughout, except for Crowley's Ridge and OK; Jun.–Aug.

Native Americans poulticed the aromatic roots of Spikenard to treat wounds, boils, and other infections.

Shooting Star *Dodecatheon meadia* L.
Primrose Family Primulaceae

The unusual swept-back petals, white or various pastel colors, along with the 5 united stamens, give this plant a distinctive appearance and account for its common name. At the base of the 15- to 25-in. stalk is a rosette of smooth ovate leaves. Also called American Cowslip, it is primarily a plant of transitional zones along the edges of glades and prairies throughout the Ozarks; Apr.–Jun.

The rare French's Shooting Star, *D. frenchii* (Vasey) Rybd., formerly considered to be s. IL's only endemic plant species, is now known also from the Ozark Plateau of AR. It can be recognized by its leaves, which narrow at their bases into petioles; Apr.–Jun.

Biennial Gaura *Gaura longiflora* Spack [*G. biennis* L.]
Evening-primrose Family Onagraceae

Gaura (rhymes with "Laura") species are tall (3–7 ft.) roadside plants with delicate white, pink, or red flowers. Each flower includes 4 petals above and 8 stamens below; the 4-branched stigma is characteristic of the family.

Biennial Gaura, also called Pink Butterfly Weed, is 3–4 ft. tall and has flowers ½–1 in. across that open in late afternoon. It is generally common throughout the Ozarks (less so on Crowley's Ridge), where it is found in prairies, glades, and dry woodlands, as well as along roadsides; Jun.–Oct.

Demaree's Gaura, *G. demareei* Raven & Gregory, is a taller plant with larger flowers that open in the morning. It occurs primarily in the Ouachita Mountains of AR; Jun.–Sep. Scarlet Gaura, *G. coccinea* Pursh, is a common smaller plant of dry soils of OK. It has reddish flowers that turn scarlet; May–Aug.

Spikenard
Aralia racemosa

Shooting Star
Dodecatheon meadia

Biennial Gaura
Gaura longiflora

Poison Hemlock *Conium maculata* L.
Carrot Family Apiaceae

Plants of the Apiaceae (formerly Umbelliferae) can be generally recognized by their tiny flowers arranged into umbels or compound umbels. Leaves are typically finely dissected like those of the cultivated carrot.

Like many other weeds, this tall (to 8 ft.), coarse biennial was introduced from Eurasia. The purple spots on the blue gray stem account for the name *maculata* (spotted). It is common in open, disturbed sites throughout most of the Ozarks (not reported from Crowley's Ridge or OK); May–Aug.

As the sap of the plant is poisonous, it was used in ancient Greece to execute criminals, including Socrates.

Spotted Cowbane *Cicuta maculata* L.
Carrot Family Apiaceae

This native biennial, which grows to 6 ft. or more in height, has leaves divided into 3–5 slender, sharply pointed leaflets. The thick stems are usually spotted with purple. Umbels may be flat-topped or round (as seen here). Also called Water Hemlock, it is restricted to moist meadows, swamps, and other wet places; it is found throughout our area; May–Sep.

Cowbane, *Oxypolis rigidior* (L.) Coult. & Rose, is a similar plant also found in wet places. It stems are green (unspotted) and its dark green leaves larger and more palmlike; Jul.–Sep.

Hairy Angelica *Angelica venenosa* (Greenway) Fern.
Carrot Family Apiaceae

Angelicas are tall, conspicuous plants with large compound umbels and a thick stem. Plants of this species reach 4–5 ft. in height and have stems streaked with purple. Leaves are divided into 3 toothed leaflets, which are, in turn, divided into smaller parts. It is a plant of moist woods and roadsides of the Ozark Plateau (MO and AR) and IL Ozarks; Jun.–Aug.

Queen Anne's Lace *Daucus carota* L.
Carrot Family Apiaceae

The bristly biennial has flat-topped inflorescences of tiny lacy florets, white or pinkish and sometimes with a single dark purple one near the center. Finely divided fernlike leaves grow from the base of the 2- to 4-ft. stems. Note the older flower cluster that has formed the "bird's nest" (upper right). It is a very common roadside weed throughout our area; May–Oct.

While young, tender roots of Wild Carrot are edible, one should make certain that they are not confused with those of other, poisonous members of this family. A root tea has been used traditionally as a diuretic and to expel worms; seeds as a "morning after" contraceptive.

Poison Hemlock
Conium maculata

Spotted Cowbane
Cicuta maculata

Hairy Angelica
Angelica venenosa

Queen Anne's Lace
Daucus carota

Sweet Cicely *Osmorhiza claytoni* (Michx.) Clarke
Carrot Family Apiaceae

Note the division of the leaves of this 2- to 3-ft. hairy plant into small toothed segments. All parts of the plant, especially the roots, release a faint aroma of anise or licorice when crushed. It occupies rich woods and thickets, mainly in uplands of MO; Apr.–Jun.

Sweet Anise, *O. longistylis* (Torr.) DC., is a similar but usually less hairy plant with a stronger aroma. It is found throughout the Ozarks; Jun.–Aug.

Native Americans used a root tea of both species of *Osmorhiza* to treat sore throats and as a panacea.

White Milkweed *Asclepias variegata* L.
Milkweed Family Asclepiadaceae

Milkweeds are recognized by their milky sap and unique flowers in umbels (see pink section for details).

White Milkweed is a tall (2–3 ft.), mostly smooth plant with ovate, opposite leaves. Margins of leaves may be entire or wavy (as seen here). It is a plant of dry to moist woods throughout our area, except for the Ozark Plateau of MO; most abundant on Crowley's Ridge; May–Jul.

Whorled Milkweed, *A. verticillata* L., is a smaller plant with whorled linear leaves and smaller umbels of white flowers arranged along the stem; May–Sep.

Other milkweeds are featured in the pink, orange, and green/brown sections.

Climbing Milkweed *Matelea baldwyniana* (Sweet) Woodson [*Gonobolus baldwyniana* Sweet]
Milkweed Family Asclepiadaceae

Matelea species are climbing or sprawling vines with 5-petaled flowers and opposite cordate leaves. Like milkweeds, climbing milkweeds produce fruits (seedpods) with plumed seeds.

Climbing Milkweed is a hairy plant that has large (6 in. long) leaves with long petioles. It occurs in rocky woods and thickets of the Ozark Plateau and Ouachita Mountains (especially along the w. edge of both regions), and occasionally on Crowley's Ridge; May, Jun.

Also called Climbing Milkweed, *M. decipiens* (Alex.) Woodson, is quite similar but has reddish brown flowers. It is found in all regions of the Ozarks except for IL Ozarks; May, Jun. Angle-pod, *M. gonocarpos* (Walt.) Shinners, has narrower leaves and greenish yellow flowers. Its common name refers to its large seedpods, which have 5 distinct ridges; Apr.–Jun.

Sweet Cicely
Osmorhiza claytoni

White Milkweed
Asclepias variegata

Climbing Milkweed
Matelea baldwyniana

Indian Hemp *Apocynum cannabinum* L.
Dogbane Family Apocyanaceae

Apocynum species, related to milkweeds, also have a milky sap and pods with plumed seeds.

Indian Hemp is a smooth, shrublike plant (to 1 ft. tall) with reddish stems. Flowers are small (less than ¼ in. wide) and bell shaped, with 5 lobes. Slender seedpods are 4–6 in. long. It is common in open, rocky soils such as prairies, glades, and along roadsides throughout the Ozarks (less common in OK than elsewhere); May–Sep.

Fibers from stems of Indian Hemp are known to have been used by Native Americans to make ropes and related items.

Spreading Dogbane, *A. androsaemifolium* L., has less erect leaves; its white flowers are pink to red inside; May–Jul.

Synandra *Synandra hispidula* (Michx.) Britt.
Mint Family Lamiaceae

Plants of the mint family are herbs, typically with square stems, opposite leaves, and a mintlike aroma when crushed. Each of the small flowers includes a corolla tube (fused petals) with 2 flaring lips: upper one 2-lobed; lower one, 3-lobed.

Synandra, a typical though uncommon mint, grows 6–12 in. tall. Flowers have white petals tinged with yellow. A biennial, it is more often seen in the s. Appalachians, but its range extends w. to the Ozarks of IL. It occupies deep rich woods, often along stream banks; May.

White Bergamot *Monarda clinopodia* L.
Mint Family Lamiaceae

Bergamots, *Monarda* species, are large coarse mints; they have whorls of flowers arranged around a terminal head with prominent bracts beneath.

White Bergamot is a smooth plant about 2 ft. tall. Both its flowers and its bracts are white or slightly tinted with purple. More common e. of the Mississippi River, it is known in the Ozarks only in Butler Co. of se. MO. It occurs along streams and in moist woods; Jun., Jul.

A pleasant tea can be made by steeping leaves and flower heads of bergamots in hot water.

Indian Hemp
Apocynum cannabinum

Synandra
Synandra hispidula

White Bergamot
Monarda clinopodia

Foxglove Beardtongue *Penstemon digitalis* Nutt.
Snapdragon Family Scrophulariaceae

Flowers of plants of the snapdragon or figwort family resemble those of the mint family (previous page). However, these plants lack the square stems and minty aroma. "Beardtongue" refers to the single bearded sterile stamen that protrudes from each flower.

This species, which grows to 4 ft. tall, has 5 conspicuous petals at right angles to the corolla tube. It is found throughout the Ozarks, where it occupies a variety of open habitats; May–Jul.

Tubed Beardtongue, *P. tubiflorus* Nutt., is a similar plant with slightly smaller flowers arranged in several tiers and closer to the stem. It is seen in similar habitats throughout the Ozarks; May, Jun.

Pale Beardtongue *Penstemon pallidus* Small
[*P. arkansanus* var. *pubescens*]
Snapdragon Family Scrophulariaceae

Plants of this species are more hairy and somewhat smaller (2 ft. tall) as compared with Foxglove Beardtongue (above). Each flower is about 1 in. long and is marked inside the corolla tube with fine purplish lines. It is a common plant of rocky woods and glades as well as roadsides throughout the Ozarks (except for the Ouachita Mountains, where it is rare or absent); May–Jul.

Arkansas Beardtongue, *P. arkansanus* Pennell, has purplish stems and clustered, darker green leaves. It occurs throughout our area except for OK and the n. Ozark Plateau of MO; Apr.–Jun.

Culver's-root *Veronicastrum virginicum* (L.) Farw.
Snapdragon Family Scrophulariaceae

Uncommon but conspicuous when present, this tall (to 7 ft.) plant produces several long spikes of white (or purplish) tubular flowers; each flower has two projecting stamens. Note the sharply pointed leaves in whorls of 3–7 per node. It inhabits open woods, thickets, and other habitats sporadically throughout the Ozarks; Jun.–Sep.

Beaked Corn-salad *Valerianella radiata* (L.) Dufr.
Valerian Family Valerianaceae

Valerianella species are generally low, forked plants with tiny 5-petaled white or bluish flowers in flat, square, or rectangular clusters. A close examination of their tiny fruits is often necessary for certain identification to species.

Beaked Corn-salad is a highly branched, 1- to 2-ft.-tall plant with clasping, opposite, sessile leaves. The stems often branch just above the attachment of the leaves. It occupies moist sites throughout the Ozarks; Apr.–Jun. "Beaked" refers to the narrow groove that runs along the fruit.

Foxglove Beardtongue
Penstemon digitalis

Pale Beardtongue
Penstemon pallidus

Culver's-root
Veronicastrum virginicum

Beaked Corn-salad
Valerianella radiata

Ox-eye Daisy *Chrysanthemum leucanthemum* L.
Aster Family Asteraceae

In the aster (also called the daisy, sunflower, or composite) family, tiny flowers (florets), each with the usual floral parts, are arranged on a common receptacle to form an inflorescence called a head. In temperate regions such as the Ozarks, this family includes the largest number of species. In late summer and fall, more than half of wildflowers encountered are likely to be of this family.

Ox-eye daisy has large flower heads that are 2 in. across. The prominent depression in the yellow disk separates it from other daisylike plants with white/yellow heads. Growing to 3 ft. tall, this Eurasian weed is found along roadsides and in fields, where it can be a serious agricultural pest; throughout our area; May–Nov.

White Snakeroot *Eupatorium rugosum* Houtt.
Aster Family Asteraceae

Eupatorium species are tall plants with opposite or whorled leaves. Clusters of fuzzy white, pinkish, or purplish flower heads appear in late summer.

White Snakeroot is a variable perennial with toothed, opposite, cordate leaves. Growing to 5 ft. in height, it is found along the edge of woods and in other semishaded sites throughout our area (less common in OK); Jul.–Oct.

In the 19th century "milk sickness" was a common and sometimes fatal disease. It was caused by drinking milk from cows that had grazed on this plant. (It is said that President Lincoln's mother died from this poisoning.) Native Americans used the roots to treat various ailments including snakebite, thus the name Snakeroot.

Other eupatoriums are featured in the blue/purple section.

Common Yarrow *Achillea millefolium* L.
Aster Family Asteraceae

The small (½ in. across) white heads of Common Yarrow are arranged into flat-topped clusters. Note the lacy, dissected leaves. This aromatic European weed is scattered across N. Amer., including all parts of the Ozarks, in open waste places; May–Nov.

In ancient times, Common Yarrow was used to treat battle wounds. Both Native Americans and European settlers used it medicinally to treat ailments such as anorexia, digestive disturbances, colds, and influenza. Recent scientific studies have documented the presence of scores of physiologically active compounds. Cultivated ornamental yarrows are available in several pastel colors.

Ox-eye Daisy
Chrysanthemum leucanthemum

White Snakeroot
Eupatorium rugosum

Common Yarrow
Achillea millefolium

Robin's Plantain *Erigeron pulchellus* Mich.
Aster Family Asteraceae

Fleabanes (*Erigeron* species) are downy, often weedy herbs with alternate, sessile leaves. Flower heads have numerous, very narrow rays.

This hairy plant, usually less than 2 ft. tall, has showy flower heads; rays may be white or vary from pale lilac to violet. Basal leaves are larger and more rounded than those on the stem above. It forms colonies in woods, especially along stream banks, throughout the Ozarks; Apr.–Jun.

Daisy Fleabane, *E. annuus* (L.) Pers., has toothed leaves with sharp points; May–Nov.

Wild Quinine *Parthenium integrifolium* L. [*P. hispidum* Raf.]
Aster Family Asteraceae

This variable smooth (or rough) perennial (5 ft. tall) has large, rough prominently toothed leaves. Tiny (¼ in. across) flower heads are arranged in umbels. Also called American Feverfew, it is found in glades, prairies, bald knobs, thickets, and other open places throughout the Ozarks; May–Sep.

Its flowering tops have been used in the treatment of malaria.

White Crown-beard *Verbesina virginica* L.
Aster Family Asteraceae

Like other members of the genus, this plant is a tall (6–7 ft.), coarse herb with "winged" stems (tissues of leaf petioles extend along the stem). Its alternate leaves are broadly lanceolate. It is distinguished from other *Verbesina* species by its flowering heads, which are composed of gray disk flowers and 3–5 white rays. It occupies open rocky places, thickets, and stream banks throughout the Ozarks; Aug.–Oct.

Wingstem, *V. alternifolia* (L.) Britt., is featured in the yellow section.

Pale Indian-plantain *Cacalia atriplicifolia* L.
Aster Family Asteraceae

All 3 Ozark species of Indian-plantain are quite similar: flat-topped clusters of white, tan, or grayish heads are held atop plants that are 3–6 ft. tall. Recognition of species depends primarily on the type of leaves.

Pale Indian-plantain is a smooth plant with white flowers and variable, but generally fan-shaped leaves with lobes or coarse teeth and whitish underneath. Though less common in the Ouachita Mountains, it is found in all portions of the Ozarks, where it occupies a variety of habitats; Jun.–Oct.

Great Indian-plantain, *C. muhlenbergii* (Sch. Bip.) Fern., has tan flowers and leaves that are similar but more rounded and green on both sides; May–Sep. Tuberous Indian-plantain, *C. plantaginea* (Raf.) Shinners, has white flowers and large oval leaves, each with 5–9 parallel veins; May–Aug.

Robin's Plantain
Erigeron pulchellus

Wild Quinine
Parthenium integrifolium

White Crown-beard
Verbesina virginica

Pale Indian-plantain
Cacalia atriplicifolia

Umbrella Magnolia *Magnolia tripetala* L.
Magnolia Family Magnoliaceae

The familiar evergreen Southern Magnolia (*Magnolia grandiflora* L.) has been reported as an occasional escape in the uplands of AR; otherwise, Ozarkian magnolias are deciduous. Like the evergreen tree, they also bear large, often fragrant flowers with white to cream-colored petals and numerous spirally arranged stamens and pistils in the center. Fruits that follow are conelike with bright red seeds.

Leaves of Umbrella Magnolia are clustered near the end of the twigs in umbrella fashion; each leaf has a long (2 ft.) tapered base. Its flowers are ill-smelling. The understory tree, which may reach a height of 30 ft., occurs primarily in rich soils of protected valleys and coves, often near small streams. In our area it is known only in the s. Ozark Plateau (AR) and Ouachita Mountains of AR and OK (single location in Le Flore Co.); Apr.–Jun; fruits, Sep.–Oct.

Cucumber Tree, *M. acuminata* L., is a larger tree with similar leaves, but they are scattered along the twigs. It occurs in all regions of the Ozarks but is less common in the s. Ouachitas and n. Ozark Plateau. It, too, is known in OK only from Le Flore Co.; Apr.–Jun.; fruits, Jul.–Sep. Bigleaf Magnolia, *M. macrophylla* Michx., known in our area only from Clay County (ne. AR) on Crowley's Ridge, has very long (to 3 ft.) leaves, their bases lobed, and silvery underneath; Apr., May; fruits, Sep.–Oct.

Virgin's-bower *Clematis virginiana* L.
Buttercup Family Ranunculaceae

Ozarkian species of *Clematis* are climbing woody vines with slender stems. Several have pink flowers and are described in the pink section.

Virgin's-bower has opposite leaves and fragrant flowers. Each flower, ¾–1 in. across, has 4 petal-like sepals and no petals. Clusters of fruits with showy silvery plumes follow. It is common in moist soils throughout (but rare in OK); Jul.–Sep.; fruits, Sep.–Dec.

Herbalists and Native Americans of various tribes have used a liniment of this and other *Clematis* species for itching and other external problems.

Silky Camellia *Stewartia malacodendron* L.
Camellia Family Theaceae

This rare, small (to 15 ft.) deciduous tree has 2- to 4-in. leaves that are softly hairy underneath. The spectacular flowers are 2–3 in. across; the pistil of each flower has 5 fused styles. Primarily a Coastal Plain species, its only known natural Ozarkian occurrence is in Ouachita County (AR) of the s. Ouachita Mountains. It grows on sandy, well-drained soils of wooded slopes; Apr.–Jul.

Umbrella Magnolia
Magnolia tripetala

Virgin's-bower
Clematis virginiana

Silky Camellia
Stewartia malacodendron

Wild Hydrangea *Hydrangea arborescens* L.
Saxifrage Family Saxifragaceae

This large (4–10 ft. tall), often straggly shrub has 3- to 6-in.-long, opposite lanceolate leaves. Small white flowers are arranged into a flat-topped or rounded corymb; there are commonly several sterile flowers around the margins of the flower cluster. It is a common shrub of rocky slopes and ravines, often along streams throughout the Ozarks; May–Sep.

Native Americans used a root tea of Wild Hydrangea to treat urinary problems; bark was poulticed on wounds and burns.

Virginia-willow *Itea virginica* L.
Saxifrage Family Saxifragaceae

This shrub, which grows to 9 ft. in height, bears numerous elongated racemes of fragrant flowers. Leaves, with their curved veins, resemble those of dogwoods (*Cornus* species) in summer but turn a much brighter red in autumn. Fruits are pointed capsules ⅜ in. long. Also called Sweetspire (an allusion to its racemes), it grows in swamps and woods and along streams throughout the uplands of AR and extending along Crowley's Ridge into se. MO; May, Jun.; fruits, Jul.–Nov.

Virginia-willow is planted as an ornamental shrub.

Gray Mock-orange *Philadelphus pubescens* Loisel.
Saxifrage Family Saxifrigaceae

Mock-oranges of the Ozarks are shrubs or small trees with simple opposite leaves and 4-petaled white flowers about an inch across. Fruit is a capsule.

Gray Mock-orange, which may grow to 15 ft. in height, has a smooth gray bark. Its numerous, fragrant flowers are clustered in groups of 5–7. It is rather common in the Ozark Plateau of AR: less so in that of MO, Ouachitas, and sw. IL; absent in OK and on Crowley's Ridge; Apr., May.

Hairy Mock-orange, *P. hirsutus* Nutt., is a smaller shrub (3–6 ft. tall) with flowers solitary or in groups of 3 or 4. In the Ozarks it is confined to the Ozark Plateau of AR. Its leaves and twigs are hairy and its trunk brownish with shredded bark; May, Jun. Scentless Mock-orange, *P. inodora* L., sometimes escapes from cultivation. In addition to having flowers that lack a fragrance, it can be separated from *P. hirsutus* by its smooth leaves and twigs; May, Jun.

Wild Hydrangea
Hydrangea arborescens

Virginia-willow
Itea virginica

Gray Mock-orange
Philadelphus pubescens

Chickasaw Plum *Prunus angustifolia* Marsh.
Rose Family Rosaceae

The genus *Prunus* includes native Ozarkian trees called cherry and peach, and some half-dozen plum species. Like other plants of the rose family, they have flowers with numerous centrally located pistils and stamens surrounded by 5 petals, usually white. The edible fruit, which contains a seed within a pit, is called a drupe. Younger bark has prominent lenticels—horizontal slits that permit gaseous exchange.

Chickasaw Plum is a small (to 10 ft. tall), highly branched tree that usually occurs in thickets as colonies connected by a common root system. Twigs and smaller branches are smooth, brownish, and zigzag, and have numerous sharply pointed side twigs. It is found in fencerows, thickets, and pastures throughout the Ozarks; Mar., Apr.; fruits, Jun., Jul.

The small, spherical, red or yellow fruits are used for pies and preserves.

Cockspur Hawthorn *Crataegus crus-galli* L.
Rose Family Rosaceae

Hawthorns are gnarly shrubs or trees widespread in open or exposed throughout the Ozarks. Unlike most other rosaceous woody plants, they have long (1–5 in.) unbranched spines attached to twigs of older branches. Leaves are simple, serrate, and either lobed or unlobed. As it takes a specialist to identify to species, no attempt is made here to distinguish this species from the approximately 15 others of the Ozarks.

This hawthorn is characterized by long (to 4 in.) straight thorns. The shiny leaves are alternate and generally finely toothed along the margin of the upper half. Flowers appear only after the leaves. Fruits (haws) are a dull red. Cockspur Hawthorn, probably our most common species, occurs throughout the Ozarks in woods and other places where it often forms thickets; Apr., May; fruits, Sep.–spring.

Haws are an important food for many types of wildlife. Although not tasty raw, they make an excellent jelly or a refreshing tea.

Alabama Snow-wreath *Neviusia alabamensis* A. Gray
Rose Family Rosaceae

This spreading shrub, which grows only to about 6 ft. in height, has flowers that differ from those of most other rosaceous plants. Petals are absent; the fringelike floral parts are calyx (sepal) lobes surrounding prominent stamens. A globally rare plant, it is know to occur in scattered sites within only 4 states: AL, TN, MO, and AR (and may be no longer present in MO). All AR sites are within the s. Ozark Plateau. It occupies moist, usually heavily shaded habitats; Apr., May.

Alabama Snow-wreath apparently reproduces only by root sprouts rather than by seed. This could, of course, help to explain its rarity.

Chickasaw Plum
Prunus angustifolia

Cockspur Hawthorn
Crataegus crus-galli

Alabama Snow-wreath
Neviusia alabamensis

Deerberry *Vaccinium stamineum* L.
Heath Family Ericaceae

Known as blueberry, huckleberry, and other names, *Vaccinium* species include shrubs with simple leaves and small white or pinkish bell-shaped flowers. Like other heaths, they are usually found in acidic soils.

Also called Highbush Huckleberry, this plant (to 6 ft. tall) has white or purplish flowers with prominent orange projecting stamens. In addition to the larger oval leaves, much smaller ones are attached at the base of the flower stalks. Fruits are green or yellowish. It occupies acidic soils of glades and oak-hickory-pine forests throughout (except absent from the IL Ozarks); May, Jun.; fruits, Jul.–Oct.

Lowbush Blueberry, *V. pallidum* Ait., which is 1–2 ft. tall, lacks the small leaves at the base of flower stalks; its flowers are greenish or pink and its blackish fruits delicious; Apr., May; fruits, Jun–Aug. Farkleberry, *V. arboreum* Marsh., is a small, highly branched (to 20 ft. tall) tree. Its evergreen leaves are dark green above, lighter beneath. Flowers are similar to those of Deerberry but without prominent stamens. Berries are black and tasteless; May, Jun.; fruits, Aug.–Dec.

Common Silverbell *Halesia tetraptera* Ellis var. *monticola* (Rehd.) Reveal &. Seldin
Storax Family Styracaceae

A shrub or small tree, Common Silverbell has large (open 1 in. long), pendant, bell-shaped flowers, their petals often tinged with pink. The species name *tetraptera* refers to its 4-winged fruits that are brownish. It has been reported from MO but is now apparently restricted in the Ozarks to the Ouachitas of AR and OK and Ozark Plateau of AR. It occurs in moist soils, usually along streams; Apr., May; fruits, Aug.–Nov.

Two-wing Silverbell, *H. diptera* Ellis, a rare tree know in the Ozarks only from Lafayette Co. (AR), is distinguished by its 2-winged fruits; Apr., May.

Large-leaved Storax *Styrax grandifolia* Ait.
Storax Family Styracaceae

Styrax species, like those of *Halesia* (above), are shrubs or small trees with simple leaves and numerous, attractive white flowers. Large-leaved Storax, which may grow to 20 ft. in height, bears large (6–8 in. long) leaves on hairy twigs. Note the 5-petaled, starlike flowers in racemes. It occurs in well-drained slopes or upland woods of the Ouachita Mountains (AR only) and s. Ozark Plateau (AR only); Apr., May.

Snowbell, *S. americana* Lam., is a somewhat smaller (10 ft. tall) shrub found in the Ozarks of AR and OK, but with an extension of its range northward along Crowley's Ridge into MO; also found in IL Ozarks. Its similar flowers are arranged singly on smoother twigs; Apr.–Jun.

Deerberry
Vaccinium stamineum

Common Silverbell
Halesia tetraptera

Large-leaved Storax
Styrax grandifolia

Fringe Tree *Chionanthus virginicus* L.
Olive Family Oleaceae

Fringe Tree is a very showy shrub or small tree. Its drooping clusters of fragrant flowers, each with 5- to 6-in.-long, slender petals, account for the alternative name, Old-man's-beard. Fruits are bluish black drupes that resemble ripe olives. Leaves are simple, elliptical to obovate, and 3–8 in. long. It is a plant of limestone ledges and edges of glades and bald knobs; also seen along creeks and moist woods. More common in the Ouachitas (AR and OK) than elsewhere, its range extends northward to the Ozark Plateau of south-central MO (absent from Crowley's Ridge and the IL Ozarks); Apr., May.

This attractive shrub deserves to be planted more widely.

New Jersey Tea *Ceanothus americanus* L.
Buckthorn Family Rhamnaceae

This low-growing shrub (to 4 ft.), also called Wild Snowball, bears complex clusters of tiny, 5-petaled flowers in the axils of its 2- to 4-in.-long, serrate, opposite leaves. It is a common plant of open, often dry, rocky places throughout our area; May–Nov.

Its common name reflects the use of its leaves in e. U.S. as a tea substitute during the Revolutionary War.

Redroot, or Inland New Jersey Tea, *C. herbaceus* Raf., has leaves that are narrower and more pointed; its flower clusters are fan shaped rather than elongated. It occurs along the w. part of the Ouachitas of AR, and Ozark Plateau (MO and AR); Apr.–Jun.

Black Locust *Robinia pseudo-acacia* L.
Pea Family Fabaceae

The showy racemes (4 to 8 in. long) of fragrant, white, pealike flowers appear just before the compound leaves, with their 3–10 pairs of ovate leaflets, unfold. Black Locust is a medium-sized, often asymmetrical tree with short, paired thorns on the branches. Common in thickets and waste places, it is found throughout the Ozarks; May, Jun.

Black Locust is a fast growing but short-lived tree. Its wood is very strong and durable, making it useful in making fence posts, railroad ties, tool handles, and other items.

Bristly Locust, *R. hispidia* L., is a showy, bristly shrub (6–10 ft. tall) with panicles of similar but rose-colored flowers; its twigs are covered with reddish brown hairs. It is native to e. U.S.; May, Jun. Yellowwood, *Cladrastis lutea* (Michx. f.) K. Koch, is a tree that occurs along the w. portion of the Ouachitas (AR and OK) and Ozark Plateau (MO, AR, and OK). Its racemes of white flowers are similar to those of Black Locust, but it is thornless and its leaves have fewer, wider leaflets; May, Jun.

Fringe Tree
Chionanthus virginicus

New Jersey Tea
Ceanothus americanus

Black Locust
Robinia pseudo-acacia

Southern Catalpa *Catalpa bignonioides* Walt.
Trumpet Creeper Family Bignoniaceae

The two Ozarkian species of *Catalpa* are similar and often difficult to distinguish. Both trees are very large, with cordate leaves and showy tubular white flowers. The large, elongated fruits resemble cigars; thus both are sometimes called Cigar Tree. Not native to the Ozarks, they are seen where they have been planted or become naturalized.

Southern Catalpa, native to the lowlands s. of the Ozarks, grows to 6 ft. in height. Its white flowers, in dense clusters, have prominent yellow and purplish markings; Jun., Jul.

Several parts of the tree have been used medicinally. A seed tea was used externally for wounds and internally for respiratory problems. Leaves were poulticed on wounds. A bark tea was considered to have antiseptic properties.

Northern Catalpa, *C. speciosa* Warder, is a taller (to 100 ft.) tree; its flowers, in more open clusters, have less prominent markings; May, Jun.

Hercules' Club *Aralia spinosa* L.
Ginseng Family Araliaceae

Also called Devil's-walkingstick, this shrub or small tree has huge doubly compound leaves up to 4 ft. long and 3 ft. wide. Small white flowers are in complex terminal clusters to 4 ft. long. Small spines are present on branches, as well as larger ones on the trunk (accounting for both common names). This distinctive plant occurs in moist soils, often along the edge of woods, throughout most of the Ozarks (absent from Ozark Plateau of OK); Jul., Aug.

The purple berries are eaten by birds. The plant is used in folk medicine for the treatment of a variety of ailments.

Poison Ivy *Toxicodendron radicans* (L.) Kuntze [*Rhus radicans* L.]
Cashew Family Anacardiaceae

This highly variable plant may trail on the ground or climb high into trees. Note the glossy leaves (bright red in the fall) with their 3 leaflets and small, white flowers. The white berries are eaten by birds, apparently with no ill effects. It grows best in moist, semishaded sites, but it occupies a variety of habitats, especially in neutral limestone soils; common throughout; May–Jul.; fruits, Jul.–Nov.

At least 50 percent of people will develop a painful dermatitis if exposed to the active ingredient, toxicodendrol, found in all parts of the plant.

Poison Oak, *T. toxicarium* (Salisb.) Gillis, also has trifoliate leaves, but the leaflets are lobed; also, it is more often an upright shrub of dry places; May. To learn more about Poison Ivy and its relatives, see Susan C. Hauser's delightful little book *Nature's Revenge*.

Southern Catalpa
Catalpa bignonioides

Hercules' Club
Aralia spinosa

Poison Ivy
Toxicodendron radicans

Common Elderberry *Sambucus canadensis* L.
Honeysuckle Family Caprifoliaceae

This common shrub grows to a height of 6–8 ft. Its showy, flat-topped flower clusters include tiny, white flowers. Large leaves are divided into 5–11 coarse-toothed leaflets. Elderberry occupies a variety of generally open sites throughout the Ozarks; May–Jul.; fruits, Aug.–Oct.

The flowers can be used to prepare fritters; the ripe purple fruits, to make jelly or wine. However, the stems and green fruits are poisonous. Native Americans used a tea made from the inner bark and also a leaf poultice to treat a variety of ailments.

Japanese Honeysuckle *Lonicera japonica* Thunb.
Honeysuckle Family Caprifoliaceae

This common vine hardly needs a description. The sweet-smelling white flowers turn cream as they mature. Introduced from Asia and naturalized throughout almost all of central and e. U.S., it is considered a noxious weed because it spreads aggressively, smothering native vegetation. It occurs throughout the Ozarks in disturbed areas and ditches, and along railroads; May, Jun.

Japanese Honeysuckle is used medicinally in Japan and China.

Yellow Honeysuckle, *L. dioica* L., native to the Ozarks, has bright yellow flowers that turn reddish as they mature, and opposite, sessile, ovate leaves; Apr., May. Trumpet Honeysuckle, *L. sempervirens* L., is featured in the red section.

American Holly *Ilex opaca* Ait.
Holly Family Aquifoliaceae

Hollies (*Ilex* species) are deciduous or evergreen shrubs or small trees. They are dioecious: the staminate (male) and pistillate (female) flowers are produced on separate plants. All Ozarkian hollies produce bright red berries.

The bark of American Holly, a medium-sized evergreen tree (to 40 ft.) is smooth and gray. The dark green leaves are thick and leathery, and have prickly margins. Primarily a component of Coastal Plain forests, its range extends n. into the Ouachitas (AR, OK) and along Crowley's Ridge of AR into se. MO (reported also from Union Co. of sw. IL). It is often planted n. of its natural range; May, Jun.; fruits, Jun.–winter.

The pale, very hard wood of this tree is useful for furniture, art work, and other specialty items.

Yaupon, *I. vomitoria* Ait., is a smaller (to 20 ft.) evergreen shrub or tree. Its elliptical leaves may be toothed but lack prickles. Its bark is light gray and smooth. Berries are red and translucent. Yaupon often forms dense thickets in the Ouachitas (AR, less often in OK); Mar.–May; fruits, Jun.–winter. Its leaves were used by Native Americans to prepare a bitter tea consumed in ritual vomiting ceremonies.

Common Elderberry
Sambucus canadensis

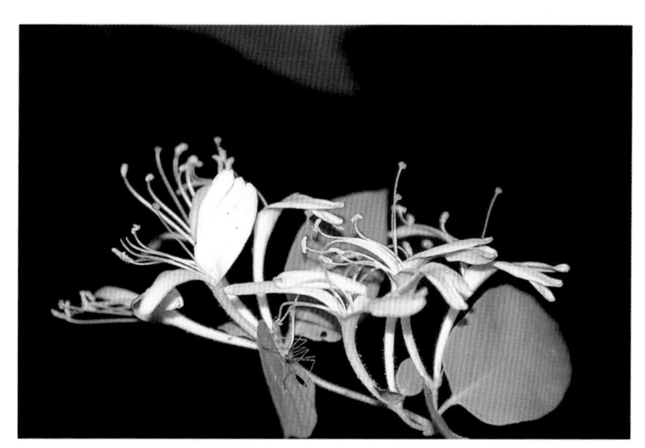

Japanese Honeysuckle
Lonicera japonica

American Holly
Ilex opaca

Winterberry *Ilex verticillata* (L.) Gray
Holly Family Aquifoliaceae

The leaves (2–4 in. long) of this deciduous shrub or small tree (to 25 ft.) are variable, often appearing as seen here but sometimes with dentate margins. Small white flowers (and red berries) are on short (½ in. or less) stalks. Winterberry is widely scattered in moist to wet sites throughout the Ozarks (except OK); Apr., May; fruits, Oct.–Jan.

Possum Haw, *I. decidua* Walt., is the most common and widespread of our Ozarkian hollies. Its leaves are more bluntly tipped; its red berries are similar but are in larger clusters of 5–6; Apr., May; fruits, Sep.–spring.

Hollies in fruit are featured in the red/orange section.

Stiff Dogwood *Cornus foemina* P. Mill. subsp. *foemina* [*C. stricta* Lam.]
Dogwood Family Cornaceae

Dogwoods are deciduous shrubs or small trees with simple leaves and clusters of small, inconspicuous greenish or white flowers. One Ozarkian species, Flowering Dogwood (below), has showy bracts (modified leaves) that surround the flower clusters. Nonbracted dogwoods are less showy.

This is probably the most common of the 5 species of nonbracted dogwoods of the Ozarks. Note its opposite leaves, white flowers in tight clusters, and reddish twigs. Fruits are dark blue berries. Its range extends from its center in the Ouachita Mountains of AR to the s. Ozark Plateau of AR and along Crowley's Ridge into MO; also in IL Ozarks. It occupies low moist sites; May, Jun.; fruits, Aug.–Oct.

Alternate-leaved Dogwood, *C. alternifolia* L. f., also has blue fruits but can be distinguished by its alternate leaves; May, Jun.; fruits, Jul.–Sep. Roughleaf Dogwood, *C. drummondii* Meyer, in addition to having leaves that are rough above, has white berries; May, Jun.; fruits, Aug.–Oct.

Flowering Dogwood *Cornus florida* L.
Dogwood Family Cornaceae

This, the only bracted dogwood of the Ozarks, is the familiar and showy native dogwood, also often planted as an ornamental. The horizontal branches bear alternate, simple leaves and inflorescences that consist of a cluster of inconspicuous greenish yellow flowers surrounded by 4 large white (or pink) notched bracts. Fruits are bright red, elongated berries. Flowering Dogwood is a common understory tree (to 35 ft.) of dry, acidic soils of oak-hickory and oak-pine forests throughout; Mar.–Jun.; fruits, Aug.–Nov.

Flowering Dogwood in fruit is featured in the red/orange section.

Winterberry
Ilex verticillata

Stiff Dogwood
Cornus foemina

Flowering Dogwood
Cornus florida

Yellow

Yellow-eyed Grass *Xyris difformis* Chapman
Yellow-eyed Grass Family Xyridaceae

Of the approximately two dozen *Xyris* species of the U.S., 5 occur in the Ozark region. All are rare to uncommon. They are plants with linear, often terete, leaves. Small 3-petaled flowers, invariably yellow, are borne on stiff, leafless stalks. Underneath the flowers are leathery scales that resemble tiny pine cones. Yellow-eyed grasses are found in roadside ditches, lakeshores, and other moist to wet situations. Identification of species often requires close examination of the seeds.

Xyris difformis, with flower spikes ½ in. long on stalks 2–3 ft. high, is known in our area only from the Ouachita Mountains of AR; Jul.–Oct.

Yellow Star Grass *Hypoxis hirsuta* (L.) Coville
Amaryllis Family Amaryllidaceae

The 6- to 12-in. grasslike leaves of this hairy perennial are longer than the stems that bear the small cluster of ½-in.-wide starlike flowers. Each flower has 3 yellow petal-like sepals, 3 yellow petals, and 6 stamens. It is common and found throughout our area in prairies, glades, and woods, both dry and wet; Apr., May (often again in fall).

White Star Grass, *H. longii* Fern., which has smaller solitary white flowers, may be seen along the s. edge of the Ouachita Mountains; Jun.–Aug.

Yellow Dog-tooth Violet *Erythronium rostratum* Wolf
Lily Family Liliaceae

The mottled leaves, together with the 6 yellow tepals, help to identify this 4- to 10-in.-tall plant. One of the first wildflowers to bloom, it is quite showy in large colonies along streams of rich woods and on bluffs throughout the Ozarks; Mar.–May.

These plants of the Ozarks were previously considered to be the Trout-lily, *E. americanum* Ker., of e. U.S. The distinction between the two species is based on chromosome number differences rather than easily observable features.

Two Ozarkian *Erythronium* species have white flowers. White Dog-tooth Violet, *E. albidum* Nutt., found in moist woodland soils, has mottled leaves and reflexed tepals; Mar.–May. Prairie Trout-lily, *E. mesochoreum* Knerr, a tallgrass prairie plant, has folded, usually unmottled leaves, and spreading (but not reflexed) tepals; Mar., Apr.

Common names used for *Erythronium* species may be of interest. "Dog-tooth" reflects the shape of the corms, bulblike underground stems (used by Native Americans to make a tea used for fevers). "Trout-lily" is apparently used because fishing for trout is often best when the plants are in flower.

Yellow-eyed Grass
Xyris difformis

Yellow Star Grass
Hypoxis hirsuta

Yellow Dog-tooth Violet
Erythronium rostratum

Large-flowered Bellwort *Uvularia grandiflora* J. E. Smith
Lily Family Liliaceae

Note the relatively large (¾–1 in. long) flowers of this stout, 1- to 1½-ft. plant; also, the clasping leaves, which are hairy underneath. It occurs in rich or rocky woods throughout the Ozarks; Apr., May.

The following are similar but smaller plants with smaller flowers: Bellwort, *U. perfoliata* L., has perfoliate leaves that are smooth underneath, May, Jun.; Small Bellwort, *U. sessilifolia* L., leaves sessile, not clasping or perfoliate; Apr., May.

Large Yellow Lady's-slipper *Cypripedium calceolus* L. var. *pubescens* (Willd.) Correll [*C. parviflorum* var. *pubescens* (Willd.) Knight]
Orchid Family Orchidaceae

Showiest of our orchids are the lady's-slippers or moccasin flowers, *Cypripedium* species.

Large Yellow Lady's-slipper is a hairy plant that grows to 2½ ft. tall. Its yellow pouch is usually 1½–3 in. long. Sepals and lateral petals are greenish to greenish brown. It is scattered throughout the Ozark Plateau and is found also in the IL Ozarks and Ouachitas of OK. Its usual habitat is protected oak-hickory forests, where it occupies acidic soils, typically on n.- or e.-facing slopes; Apr.–Jun.

Small Yellow Lady's-slipper, *C. calceolus* L. var. *parviflorum* (Salisb.) Fern., is a smaller, smooth plant (to 8 in.) with a smaller yellow pouch (less than 1½ in. long) and more tightly twisted sepals and lateral petals; Apr.–Jun. Southern Lady's-slipper, *C. kentuckiense* Reed, also has a yellow pouch, but maroon-brown sepals and lateral petals; Apr.–Jun.

Yellow Fringed Orchid *Platanthera ciliaris* (L.) Lindley [*Habenaria ciliaris* (L.) R. Br.]
Orchid Family Orchidaceae

The exquisite flowers of this 2-ft.-tall plant vary from yellow to deep orange. Note the long spurs that extend from the lips of each flower. This rare orchid occurs in wet, acidic soils along springs, near sinkholes, or in moist pine woods. It is known from all major regions of the Ozarks (except absent from OK and IL) but is seen most often in the Ouachita Mountains of AR; Jul.–Oct.

Crested Fringed Orchid, *P. cristata* (Michx.) Lindley, of the Ouachita Mountains of AR, has smaller, yellow flowers with much shorter spurs; Jun.–Sep. These two species are known to hybridize in the Appalachians and may also do so in the Ozarks.

These and other members of the orchid family deserve our protection. As they do not transplant well, one should not collect them from the wild. Purple Fringeless Orchid, *P. peramoena* Gray, is featured in the blue/purple section.

Large-flowered Bellwort
Uvularia grandiflora

Large Yellow Lady's-slipper
Cypripedium calceolus

Yellow Fringed Orchid
Platanthera ciliaris

American Lotus *Nelumbo lutea* (Willd.) Pers.
Water-lily Family Nymphaceae

This quite unmistakable plant has huge (1–2 ft. across), blue-green, bowl-shaped leaves held well above the water by smooth petioles. Each of the large (6–8 in. across) flowers consists of numerous pale yellow petals surrounding the saltshaker-like pistil. Each fruit (lower left) contains seeds noted for their ability to germinate decades after maturing. It occurs in shallow water of lakes and sloughs throughout our area (less common in the Ouachitas); Jun.–Sep.

The roots, leaves, and seeds of the plant, which was called Yanquapin by Native Americans, were used for food. It is also important as a shelter for fish and wildlife.

Spatterdock *Nuphar lutea* (L.) J.E. Smith
Water-lily Family Nymphaceae

As compared with American Lotus (above), this plant has floating, deeply notched leaves and much smaller (1–2 in. across) flowers. Sepals are greenish yellow; petals, bright yellow. It occurs in shallow water of streams, ponds, and sloughs throughout our area except for OK and the IL Ozarks; May–Oct.

Hooked Buttercup *Ranunculus recurvatus* Poir.
Buttercup Family Ranunculaceae

Nearly 20 species of buttercups (*Ranunculus* species) are found in the Ozarks. Herbs with deeply cut, alternate leaves, they inhabit dry to wet, open places. Flowers have varying numbers of yellow (or white) sepals and petals and numerous bushy stamens. Buttercups are generally poisonous if ingested; the sap may also be a skin irritant.

This common 1- to 2-ft.-tall buttercup is a hairy plant with inconspicuous flowers. The green, recurved sepals are longer than the small yellow petals. "Hooked" refers to the beaks on the achenes (small, single-seeded fruits). It is a common plant throughout our area. It occupies moist sites along the edges of springs, streams, and in moist woods; May–Jul.

These buttercups are also widespread species with small yellow flowers but smaller leaves. Kidney-leaf Buttercup, *R. abortivus* L., has small, fan-shaped basal leaves with crenated margins; Mar.–Jun. Leaves of Early Buttercup, *R. fascicularis* Muhl., are divided into narrow, branched segments; Mar.–May.

These buttercups grow submerged in water; their leaves are subdivided into narrow filaments: Yellow Water-buttercup, *R. flabellaris* Raf., May, Jun.; and White Water Crowfoot, *R. longirostris* Godron, May–Sep.

American Lotus
Nelumbo lutea

Spatterdock
Nuphar lutea

Hooked Buttercup
Ranunculus recurvatus

Celandine Poppy *Stylophorum diphyllum* (Michx.) Nutt.
Poppy Family Papaveraceae

Below each cluster of yellow-petaled flowers (1½–2 in. across) is a single pair of large, deeply dissected leaves with rounded lobes. Other leaves are basal. Seed pods are ovoid and hairy. Also called Wood Poppy, it is found sporadically in rich, moist soils of the Ozark Plateau of AR and MO, Crowley's Ridge of MO, and the IL Ozarks; Apr.–Jun.

The alien Celandine, *Chelidonium majus* L., of the same family presents a similar appearance. However, its leaves are attached singly to the stem, its flowers (¾ in. across) are smaller, and its fruit is a smooth, slender seedpod; Apr.–Sep. The caustic sap of Celandine has been used to treat warts.

Pale Corydalis *Corydalis flavula* (Raf.) DC.
Fumitory Family Fumiaraceae

Closely related to Dutchman's-breeches (white section), *Corydalis* species are small plants with finely cut leaves and narrow, tubular yellow flowers about ½ in. long.

Also called Yellow Corydalis, this plant has pale yellow flowers, each with a crest on its upper, toothed petal. It occurs in rich woods, especially along streams, in ravines, and on moist ledges along bluffs, throughout; Apr., May.

Small-flowered Corydalis, *C. micrantha* (Engelm.) Gray, also widespread, has an untoothed crest; May, Jun. Golden Corydalis, *C. aurea* Willd., rare on the Ozark Plateau of MO, lacks the crest; its leaves are silvery green; Mar.–Jun.

Rough-fruited Cinquefoil *Potentilla recta* L.
Rose Family Rosaceae

Cinquefoils are small to medium-sized plants, generally with yellow flowers. Like other members of the rose family, their flowers have 5 petals. "Cinquefoil" means "5-leaved"; in reality, they have compound leaves divided into 3 or more leaflets, according to the species.

Rough-fruited Cinquefoil is a tall (to 2 ft.) plant with 5–7 toothed, relatively narrow leaflets. Another name, Sulfur Cinquefoil, indicates the pale yellow of the petals. A European plant, it occurs along roadsides and other disturbed places throughout much of the Ozarks; May–Aug.

Tall Cinquefoil, *P. arguta* Pursh, is a native plant similar to *P. recta* but has white to creamy flowers; May–Aug.

Other cinquefoils have flowers with bright yellow flowers. Common Cinquefoil, *P. simplex* Michx., is a low, trailing plant. Its leaves consist of 5 wider leaflets with teeth that are more prominent; Apr.–Jun. The less common Rough Cinquefoil, *P. norvegica* L., is an upright plant. Its leaves consist of 3 leaflets; May–Oct.

Celandine Poppy
Stylophorum diphyllum

Pale Corydalis
Corydalis flavula

Rough-fruited Cinquefoil
Potentilla recta

Creamy Wild Indigo *Baptisia bracteata* Nutt. [*B. leucophaea* Nutt.]
Pea Family Fabaceae

This bushy plant, which has hairy, spreading branches, grows to 2 ft. tall. Leaves are divided into 3 leaflets, but a pair of bracts at their bases gives the appearance of 5 leaflets. Numerous pealike flowers cause the branches to droop. It occupies glades, prairies, and rocky roadsides throughout the Ozarks (less common on Crowley's Ridge and sw. IL); Apr.–Jun.

Pawnees used root and leaf teas of *B. bracteata* to treat a variety of ailments. Modern research indicates that its extracts can stimulate the immune system and thus validates some of these uses.

Other *Baptisia* species are featured in the white and blue/purple sections.

Pencil Flower *Stylosanthes biflora* (L.) B.S.P.
Pea Family Fabaceae

Pencil Flower is a small, hairy plant with wiry, branched stems that often bend or trail over the ground. Note the large standard upright petal of the flower and trifoliate leaves. Flowers are ¼ in. long and may be yellow-orange. The only Ozarkian species of the genus, it occurs throughout, especially in acidic soils of glades, prairies, and open woods; May–Sep.

Mohlenbrock and Voight, in *A Flora of Southern Illinois*, note that there is great variation in degree of hairiness among individuals of Pencil Flower in sw. IL.

Partridge Pea *Chamaecrista fasciculata* (Michx.) Greene [*Cassia fasciculata* Michx.]
Caesalpinia Family Caesalpiniaceae

Historically included in the pea family (Fabaceae), plants of the senna family have, like those of the pea family, compound leaves and podlike fruits, but their flowers have more nearly equal petals.

Partridge Pea is a weedy annual that grows to 2 ft. in height. Its flowers, arranged along the stem, typically have some purplish markings at the base of the petals. It occurs, often in large colonies, in dry, sunny waste places throughout the Ozarks; Jun.–Oct.

Also common, Wild Sensitive-plant, *C. nictitans* L., is a smaller plant (6–12 in. tall) that produces smaller flowers and has leaves that fold when touched; Jul.–Sep. Wild Senna, *Senna marilandica* (L.) Link, has similar flowers arranged in clusters along its 8-ft.-tall stem; Jul.–Aug.

Creamy Wild Indigo
Baptisia bracteata

Pencil Flower
Stylosanthes biflora

Partridge Pea
Chamaecrista fasciculata

Missouri Evening-primrose *Oenothera macrocarpa* Nutt. [*O. missouriensis* Sims]

Evening-primrose Family Onagraceae

Evening-primroses (not related to primroses, Primulaceae) are herbs with a distinctive flower: 4 reflexed sepals and 4 petals attached at the end of a long calyx tube; the stigma has 4 branches that form a cross. The common name is derived from the several species that open toward evening (others, called sundrops, open earlier in the day).

This species produces huge flowers (3–4 in. across) on a plant less than 1 ft. in height. Petals turn orange as they wither. *Macrocarpa* refers to the large (2 in. long), 4-winged fruits (seed capsules). It is locally abundant on limestone glades and bald knobs of the Ozark Plateau of southcentral MO and northcentral AR; also in OK; May–Aug.

Local folk use the name Glade Lily for this spectacular plant. It is also called Ozark Sundrops. It does not transplant well but can be easily started from seed sown in the fall.

Common Evening-primrose, *O. biennis* L., is a much taller (to 5 ft.) plant with a hairy, often reddish stem. Flowers are 1–1½ in. across. It is common in disturbed, open sites throughout; Jun.–Sep. Cut-leaved Evening-primrose, *O. laciniata* Hill, is a smaller plant with incised leaves; Jun.–Sep.

Showy Evening-primrose, *O. speciosa* Nutt., is featured in the pink section.

Sundrops *Oenothera fruticosa* L.

Evening-primrose Family Onagraceae

This variable plant, 1–3 ft. tall, has narrowly lanceolate leaves with entire margins. Its day-blooming flowers, 1 in. or more across, have stamens that are often orange. The seedpods have prominent ridges. It occupies rocky or sandy open woods throughout most of our area (absent from IL Ozarks); May–Aug.

Floating Primrose-willow *Ludwigia peploides* (H.B.K.) Raven [*Jussiaea repens* Ktze.]

Evening-primrose Family Onagraceae

Also called water-primroses, *Ludwigia* species are mostly tropical plants. Note this plant with its shiny, long-stalked flowers (¾–1 in. across), each with 10 stamens and petals. The winged stems of the aquatic perennial produce roots at their lower nodes, allowing it to form dense colonies in shallow water of ponds and slow-moving streams. It occurs sporadically throughout most of the Ozarks; May–Oct.

Seedbox, *L. alternifolia* L., has a similar growth habit and also grows in wet places. Its stems are not winged and its flowers each have 4 stamens and 4 nearly round yellow petals. Its common name reflects its squarish seed capsules; Jun.–Aug.

Missouri Evening-primrose
Oenothera macrocarpa

Sundrops
Oenothera fruticosa

Floating Primrose-willow
Ludwigia peploides

Prairie Parsley *Polytaenia nuttallii* DC.
Carrot Family Apiaceae

See the white section for an overview of the carrot family.

Also called Prairie Parsley, this is a tall (to 3 ft.), branching plant with delicate umbels of pale yellow flowers. Parsleylike leaves are arranged along the length of the stem. It grows in glades and prairies along the w. edge of the Ozark Plateau (MO, AR, and OK), and also in the Ouachitas; Apr.–Jun.

The specific epithet honors its discoverer, Thomas Nuttall (1786–1859).

Meadow Parsnip *Thaspium trifoliatum* (L.) Gray
Carrot Family Apiaceae

This highly branched plant grows to 2 or 3 ft. tall. Its stems and leaves are smooth. Leaves vary in shape: basal ones are usually cordate, with those attached to the stem (seen here) divided into 3 finely toothed leaflets; the tiny, light yellow flowers (sometimes purple) are arranged into umbrella-like compound umbels. It grows in rocky, open, usually moist woods throughout except for OK; May, Jun.

Hairy-jointed Meadow Parsnip, *T. barbinode* (Michx.) Nutt., found along streams throughout, is well named: it differs from *T. trifoliatum* by having a fringe of hairs at the nodes; Apr.–Jun. For comparison, see *Zizia* species below.

Golden Alexanders *Zizia aurea* (L.) Koch
Carrot Family Apiaceae

Also called Meadow Parsnip, this plant resembles *Thaspium* species (above), but lacks the cordate basal leaves. Its leaves are typically doubly compound: each is divided into 3 parts, each further subdivided into 3–11 smaller, prominently toothed leaflets. Stems are usually reddish. Growing to 3 ft. in height, it occurs throughout the Ozarks in moist prairies, woods, and glades; Apr.–Jun.

Also called Golden Alexanders, *Z. aptera* (Gray) Fern., differs by its basal leaves which are simple (not divided) and cordate. Also, its stem leaves are compound (vs. doubly compound). It can be distinguished from Meadow Parsnip (above), which has similar leaves, by the central flower of each umblet, which is stalkless; Apr.–Jun.

Golden Alexanders were used medicinally by both Native Americans and early settlers. However, like many plants of this family, it is potentially dangerous. The genus name commemorates Johann B. Ziz, a German botanist (1779–1829).

Prairie Parsley
Polytaenia nuttallii

Meadow Parsnip
Thaspium trifoliatum

Golden Alexanders
Zizia aurea

Yellow Violet *Viola pubescens* Ait. [*V. pensylvanica* Michx.]
Violet Family Violaceae

This stemmed violet (flowers and leaves on same stalk), usually 6–19 in. tall, has yellow flowers on long petioles. Note the purplish veins near the base of each petal. Leaves are cordate with scalloped margins. Despite the species name *pubescens*, which means "hairy," plants may be either smooth or hairy. The only yellow-flowered Ozarkian violet, it occurs in moist woods throughout (except in OK); Mar.–May.

Eastern Prickly Pear *Opuntia humifusa* (Raf.) Raf. [*O. compressa* (Salisb.) Macbr.]
Cactus Family Cactaceae

This low-growing, sprawling cactus has rounded "pads" (stems) up to 5 in. long. The pads have 1 or 2 needlelike spines that emerge from each cluster of much smaller bristles. Showy flowers, to 4 in. across, have 8–10 waxy sepals and petals; often there is a reddish center, as seen here. It occurs in open glades, rocky prairies, and other dry, sunny sites sporadically throughout most of the Ozarks; May–Jul.

Western or Plains Prickly Pear, *O. macrorhiza* Engelm., can be distinguished by its pads, which have more than 2 spines per cluster. It occurs along the w. edge of the Ozarks; May–Jul.

Common St. John's-wort *Hypericum perforatum* L.
St. John's-wort Family Clusiaceae

Among St. John's-worts (*Hypericum* species) are herbs and shrubs bearing simple, opposite leaves with entire margins; leaves are often dotted with minute glands. Flowers have 5 (or 4) yellow petals and numerous stamens. The common name was given because they flower in midsummer at the time of the birthday of St. John the Baptist.

This European plant, our only alien St. John's-wort, is a highly branched, 1- to 2-ft.-tall herb that bears numerous flowers to 1 in. across. The petals have black dots along their margins. It is relatively common in the IL Ozarks and on the Ozark Plateau of MO; less so on that of AR, Ouachita Mountains, and Crowley's Ridge. It is primarily a roadside weed; May–Sep.

There is a renewed medicinal interest in St. John's-wort. Recent studies at Vanderbilt University indicate that it is marginally useful for the treatment of mild to moderate mental depression. If livestock are exposed to strong sunlight after grazing on the plant, they may develop a dermatitis.

The native Round-fruited St. John's-wort, *H. sphaerocarpum* Michx., is very similar, but plants of a colony are connected by underground runners; May–Sep. Spotted St. John's-wort, *H. punctatum* Lam., is also similar, but its flowers are smaller and it has prominent black dots on its leaves as well as on its flower petals; Jun.–Sep.

Yellow Violet
Viola pubescens

Eastern Prickly Pear
Opuntia humifusa

Common St. John's-wort
Hypericum perforatum

Fringed Loosestrife *Lysimachia ciliata* L.
Primrose Family Primulaceae

Loosestrifes are summer-flowering perennials, mostly 1–3 ft. tall, that grow in moist, shaded or semishaded woods or thickets. Their numerous starlike flowers, each with 5 yellow, pointed petals, may be terminal (clusters at top of the plant), axillary (attached at the leaf nodes), or both.

Fringed Loosestrife, which grows to 3 or 4 ft. tall, has nodding terminal and axillary flowers. There are prominent fringes on the petioles of the opposite (paired) leaves. It occurs on the Ozark Plateau (MO, AR, OK), and less commonly in the Ouachitas, Crowley's Ridge, and the IL Ozarks; May–Jul.

Yellow Coneflower *Echinacea paradoxa* (Norton) Britt.
Aster Family Asteraceae

The aster family (also called the sunflower, daisy, or composite family) is the largest one of the Ozarks (i.e., it has the most species). Members of this family are generally recognized by their characteristic inflorescence: the head is composed of many small flowers (florets) clustered together on a common receptacle (fig. 2). If you examine a single floret under a hand lens, you will see that it has a floral formula of 5-5-5-1, with 5 anthers to form a cylinder around the style; above the style is a 2-lobed stigma. In most species, disc florets make up the center, with ray florets arranged around the outer edge.

Echinacea species bear large daisylike flower heads, each consisting of a dark spiny cone surrounded by conspicuous rays, usually lavender to purple. Yellow Coneflower, the only species with yellow rays (thus the species name *paradoxa*), is a smooth plant with stems to 3 ft. tall. Basal leaves are large (to 10 in. long) and lanceolate. It occurs sporadically in limestone glades and bald knobs within the w. half of the Ozark Plateau of MO and a few counties of the n. Ozark Plateau of AR; May, Jun.

Other *Echinacea* species are featured in the blue/purple section.

Stiff Coreopsis *Coreopsis palmata* Nutt.
Aster Family Asteraceae

Called tickseeds or tick-seed sunflowers, *Coreopsis* species are herbs with 6–10 (often 8) rays per flower head; rays are yellow and often toothed at their tips. Buds are typically spherical. Stiff Coreopsis is so named because of its rigid, narrow sessile leaves that branch into 3 smaller divisions. The 1–3 ft.-tall plant grows in prairies, glades, and fields throughout the Ozarks; May–Jul.

Tall Tickseed, *C. tripteris* L., is a taller plant (to 8 ft.) with petioled leaves divided into 3 segments; Jul.–Sep. Lance-leaved Coreopsis, *C. lanceolata* L., has much larger, lanceolate leaves; Apr.–Jul.

Fringed Loosestrife
Lysimachia ciliata

Yellow Coneflower
Echinacea paradoxa

Stiff Coreopsis
Coreopsis palmata

Gray-headed Coneflower *Ratibida pinnata* (Vent.) Barnh.
Aster Family Asteraceae

Also called Drooping Coneflower, this plant is easily recognized by its long extremely reflexed rays. The cone changes from greenish to dark gray or brown as it matures. As suggested by *pinnata,* the leaves are deeply dissected. It is frequently seen in prairies, edges of woods, and other dry places of the Ozark Plateau (MO, AR, OK) and IL Ozarks; less often in the Ouachitas; May–Sep.

The less common Long-headed Coneflower, *R. columnaris* (Nutt.) Woot. & Standl., has a light gray, more elongated cone and yellow or reddish brown reflexed rays. Leaves are dissected into numerous narrow segments; Jun.–Sep.

Black-eyed Susan *Rudbeckia hirta* L.
Aster Family Asteraceae

Rudbeckia species, often called coneflowers, have brown, conical discs surrounded by long yellow rays. Black-eyed Susan is a bristly, 1- to 3-ft.-tall plant with unlobed leaves and one to several flower heads. Each head, 2 to 3 in. wide, has 10–30 rays. The plant brightens roadsides, old fields, and open woods throughout our area; May–Oct.

These plants have similar flower heads. Sweet Coneflower, *R. subtomentosa* Pursh, has lower leaves that are 3-cleft; Jul.–Oct. Missouri Black-eyed Susan, *R. missouriensis* Engelm., is a smaller, more hairy plant with lower leaves lanceolate, upper ones linear. It often dominates glades of the Ozarks; Jun.–Oct. Orange Sunflower, *R. fulgida* Ait., has a ring of orange at the base of each ray; Jul.–Oct.

Thin-leaved Coneflower, *R. triloba* L., is a more branched plant with smaller (1 in. across), more numerous heads and 3-lobed lower leaves; Jun.–Nov.

The genus commemorates Olaf Rudbeck (1630–1702), botanist and teacher of Carolus Linnaeus.

Wild Goldenglow *Rudbeckia laciniata* L.
Aster Family Asteraceae

This tall (3–12 ft.), highly branched plant has leaves divided into 3–7 sharply pointed lobes (lower leaves are more lobed than the upper ones seen here). Note also the green heads (alternative name, Green-headed Coneflower) and somewhat reflexed rays. It is common in moist soils throughout the Ozarks (except absent on Crowley's Ridge); Jul.–Sep.

Native Americans made from several *Rudbeckia* species a root tea that was used for worms, indigestion, and snakebite.

Gray-headed Coneflower
Ratibida pinnata

Black-eyed Susan
Rudbeckia hirta

Wild Goldenglow
Rudbeckia laciniata

Bearsfoot *Polymnia uvedalia* L.
Aster Family Asteraceae

Bearsfoot, also know as Yellow-flower Leafcup and Large-flowered Leafcup, has hairy, maplelike leaves with winged petioles. The flower heads, 2–3 in. across, include both yellow disc and ray florets. It occurs in rich, moist soils of woods and thickets throughout; Jul.–Sep.

These leafcups have smaller flower heads with yellow discs and white rays (sometimes absent) and unwinged stems. Common Small-flowered Leafcup, *P. canadensis* L., has large, toothed, lanceolate leaves; it is found on limestone soils; May–Oct. Heartleaf Leafcup, *P. cossatotensis* Pittman & Bates, recently discovered in the Ouachitas of AR, has large cordate leaves; Aug.–Oct.

Autumn Sneezeweed *Helenium autumnale* L.
Aster Family Asteraceae

This 2- to 5-ft. perennial has winged stems and coarsely toothed, lanceolate leaves (dried, the plant may cause sneezing). Note the wedge-shaped, slightly reflexed rays that surround the yellow-gray globular discs. It grows along streams, in swamps, and in wet meadows throughout the Ozarks but is less common in the Ouachitas and on Crowley's Ridge; Aug.–Nov.

Native Americans used dried, powdered flowers as a snuff for colds.

Purple-headed Sneezeweed, *H. flexuosum* Raf., has sparse, narrowly lanceolate leaves and reddish brown discs; Jun.–Nov. Bitterweed, *H. amarum* (Raf.) Rock, is a much shorter (to 1 ft.) plant with numerous grasslike leaves. It inhabits roadsides and overgrazed pastures; Jun.–Oct.

Helenium species contain helenalin, a lactone recently found by the National Cancer Institute to be an anticancer agent.

Western Tickseed *Bidens aristosa* (Michx.) Britt.
Aster Family Asteraceae

Bidens aristosa is 2–4 ft. tall and has large, compound leaves, each of which is divided into 5 toothed, sharply pointed leaflets. Flower heads are 1–2 in. wide and have 6–10 rays. This and other *Bidens* species have barbed achenes (small, dry, one-seeded fruits), known as "beggar ticks," that attach themselves to fur or clothing. Western Tickseed occurs in sunny, wet sites throughout the Ozarks; Aug.–Oct.

Spanish-needles, *B. bipinnata* L., is recognized by its 4-angled stems, wider, incised leaflets, and flower heads with shorter, cream-colored rays; Aug.–Oct.

Bearsfoot
Polymnia uvedalia

Autumn Sneezeweed
Helenium autumnale

Western Tickseed
Bidens aristosa

Wingstem *Verbesina alternifolia* (L.) Britt.
[*Actinomeris alternifolia* (L.) DC.]
Aster Family Asteraceae

"Wingstem" refers to the vertical "wings" on stems that are continuous with the leaf petioles. Note the alternate leaves and flower heads with green discs and 2–8 reflexed rays. The plant, also called Yellow Ironweed, grows to 7 ft. It occurs in moist woods and thickets throughout the Ozarks (but apparently absent from OK); Aug.–Oct.

Yellow Crown-beard, *V. helianthoides* Michx., has more numerous (8–15) horizontal (not reflexed) rays; May–Oct.

White Crown-beard, *V. virginica* L., is featured in the white section.

Cup Rosin-weed *Silphium perfoliatum* L.
Aster Family Asteraceae

Silphium species are tall plants with large leaves and yellow sunflower-like flower heads. The half-dozen species found in the Ozarks are distinguished primarily by their leaf shape.

Cup Rosin-weed, also called Cup-plant, has smooth, square stems, 4–8 ft. in height. Its 2 upper leaves are perfoliate; their bases surround the stem to form a shallow cup. It occupies moist sites throughout our area; Jul.–Sep.

Entire-leaved Rosin-weed, *S. integrifolium* Michx., a coarse plant with a resinous sap, has large simple, opposite, lanceolate leaves; Jul.–Sep. Compass-plant, (*S. laciniatum*) has very large (1–2 ft. long), deeply incised, mostly basal leaves; Jul.–Sep. Prairie-dock, *S. terebinthinaceum* Jacq., has huge (1–2 ft. long) cordate, mostly basal leaves; Jul.–Oct.

Woodland Sunflower *Helianthus divaricatus* L.
Aster Family Asteraceae

Sunflowers are tall plants with rough or hairy stem and leaves. Leaves vary in shape from cordate to lanceolate; most have dentate margins. Flower heads are often large and showy. Rays are yellow and wide, and usually overlap at their bases.

Woodland or Rough Sunflower, which may grow to 6 ft. tall, has thick, rough, lanceolate leaves and large flower heads. It is found in fields, open woods, often in disturbed soil throughout; Jul.–Oct.

Ashy Sunflower *Helianthus mollis* Lam.
Aster Family Asteraceae

This sunflower is characterized by very hairy grayish stems (3–4 ft.) and opposite cordate leaves, also hairy. Flower heads, each 3–4 in. across, are on long stalks. Numerous (to 30) lemon yellow rays surround the darker yellow disc. It is seen primarily in prairies, sporadically throughout the Ozarks. It is not generally common but often occurs in large stands locally; Jul.–Oct.

Wingstem
Verbesina alternifolia

Cup Rosin-weed
Silphium perfoliatum

Woodland Sunflower
Helianthus divaricatus

Ashy Sunflower
Helianthus mollis

Roundleaf Ragwort *Senecio obovatus* Muhl.
Aster Family Asteraceae

Senecio species includes plants with flat-topped heads of yellow flowers. "Ragwort" comes from the random arrangement of the yellow rays encircling the heads, giving them a ragged appearance. Roundleaf Ragwort, a small (1–1½ ft. tall) plant, has oval basal leaves that are broader above the middle than below. It is seen at the base of ledges and on the edge of glades throughout the Ozarks except for IL Ozarks; Apr.–Jun.

Two other widespread ragworts occupy more moist sites. Golden Ragwort, *S. aureus* L., is a larger plant (1–3 ft. tall) and has cordate basal leaves with long petioles; Apr.–Jun. Butterweed, *S. glabellus* Poir., has large, deeply dissected basal leaves and thick, hollow stems; May–Jul.

Potato Dandelion *Krigia dandelion* (L.) Nutt. [*Cynthia dandelion* DC.]
Aster Family Asteraceae

Krigia species resemble small dandelions. Flower heads are yellow or orange. Their leaves have jagged margins. Their sap is milky. Potato Dandelion, to 1 ft. tall, has solitary flower heads ½ in. across. Lobed, variable leaves are basal. Roots bear tubers that resemble small potatoes. It inhabits moist or rocky acidic soils throughout the Ozarks; Apr.–Jun.

Dwarf Dandelion, *K. virginica* (L.) Willd., is a similar but smaller (3 in. tall) plant that often occurs in crevices of rock outcroppings; Apr.–Aug. False Dandelion, *K. biflora* (Walt.) Blake, is a 2-ft.-tall plant with orange-yellow flower heads. Despite the species name *biflora,* the number of heads per stem is variable. Leaves include large dandelion-like basal ones and smaller ones above that clasp the stem; May–Aug.

Canadian Goldenrod *Solidago canadensis* L. [*S. altissima* L.]
Aster Family Asteraceae

Goldenrods are fall-flowering, perennial herbs with tiny yellow florets in showy clusters. Leaves are simple and alternate. Identification of the two dozen Ozarkian species is often difficult due to hybridization. Canadian Goldenrod, which may reach 6 ft., has heads clustered on arching side branches. Its parallel-veined, lanceolate leaves are rough above and hairy beneath; margins are markedly dentate. Stems are grayish, covered with down. Probably our most common goldenrod, it occurs in open places throughout; Aug.–Nov.

Two other Ozarkian species also have showy plumes and parallel-veined, lanceolate leaves. Sweet Goldenrod, *S. odora* Ait., has toothless licorice-scented leaves; Jul.–Oct. Late Goldenrod, *S. gigantea* Ait., has pale green or purplish stems and leaves less markedly toothed than those of Canadian Goldenrod; Jul.–Oct. The leaves of Sweet Goldenrod can be used to make an excellent tea.

Roundleaf Ragwort
Senecio obovatus

Potato Dandelion
Krigia dandelion

Canadian Goldenrod
Solidago canadensis

Rough-leaved Goldenrod *Solidago rugosa* Mill.
Aster Family Asteraceae

Like the goldenrods featured on the previous page, this one also has plumelike clusters of flower heads, but with shorter side branches. Its leaves are rough and wrinkled, and have deeply toothed margins; also the veins are featherlike (vs. parallel). Stems are very hairy. It occupies wet sites throughout the Ozarks (less common on Crowley's Ridge); Aug.–Oct.

Sharp-leaved Goldenrod, *S. arguta* Ait., has a smooth, usually brownish stem. Its leaves are somewhat like those of *S. rugosa* except that the leaf margins are less zagged and the blade ends in a slender, pointed tip; Aug.–Oct.

Gray Goldenrod *Solidago nemoralis* Ait.
Aster Family Asteraceae

Gray Goldenrod is a small (½–2 ft.), unbranched plant with 1-sided plumes; it is covered with fine hairs, giving it a gray-green appearance. Note also the tiny stipules (small, paired leaflike structures) at the base of its alternate leaves. Also called Old-field Goldenrod, it is found in dry, upland prairies and open woods throughout the Ozarks; Jun.–Nov.

This is one of our smallest goldenrods and also one of the most tolerant of dry conditions. This and other goldenrods have been used as dye plants; the result is an intense yellow to gold color.

Zigzag Goldenrod *Solidago flexicaulis* L.
Aster Family Asteraceae

Note the zigzag stem and tufts of flower heads in the leaf axils of this 1- to 3-ft.-tall goldenrod. The large, broad, toothed leaves account for its also being called Broad-leaved Goldenrod. It occurs in rich open and shaded woods of the Ozark Plateau (MO, AR), less often on Crowley's Ridge; Jul.–Oct.

Blue-stemmed Goldenrod *Solidago caesia* L.
Aster Family Asteraceae

The arrangement of the long, lanceolate leaves and spacing of the clusters of flower heads along the purplish stems help to identify this smooth plant. It grows to 3 ft. tall. Also called Wreath Goldenrod, it occupies bluffs and rocky woods throughout the Ozarks; Aug.–Oct.

Hairy Goldenrod, *S. hispida* Muhl. ex. Willd., also has tufts of flower heads spaced above the stem at the base of lanceolate leaves. It differs by having very hairy stems and leaves; Jul.–Oct. The rare Ouachita Goldenrod, *S. ouachitensis* C. & J. Taylor, also resembles *S. caesia* but has long, ovate leaves; Sep.–Nov.

Rough-leaved Goldenrod
Solidago rugosa

Gray Goldenrod
Solidago nemoralis

Zigzag Goldenrod
Solidago flexicaulis

Blue-stemmed Goldenrod
Solidago caesia

Tulip-tree *Liriodendron tulipifera* L.
Magnolia Family Magnoliaceae

This large (to 100 ft. tall) tree has leaves whose sharp-tipped lobes resemble those of some maples. Note the large (2 in. across) flowers, each with 3 reflexed sepals and 6 petals marked with orange. Also called Yellow Poplar (but not to be confused with true poplars, *Populus* species), it is a common and widespread commercial lumber tree e. of the Mississippi River. In our area, Tulip-tree occurs naturally only on Crowley's Ridge (AR and MO) and the IL Ozarks. It occupies moist woods, often forming dense successional stands after logging; Apr., May.

Native Americans used a bark tea internally for indigestion, worms, fevers, and coughs; externally, as a wash for snakebites, boils, and wounds.

Ozark Witch-hazel *Hamamelis vernalis* Sarg.
Witch-hazel Family Hamamelidaceae

This deciduous shrub or small tree (to 15 ft. tall) has large leaves (4 in. long) with entire to undulate margins. Each flower includes 4 twisted petals, each about ½ in. long and fragrant; color varies from yellow to orange-red. Also, called Vernal Witch-hazel it occurs in moist soils along stream banks and n.-facing slopes throughout the Ozark Plateau and Ouachitas; Jan.–Apr.

Eastern or Common Witch-hazel, *H. virginia* L., is a similar but more widespread shrub of e. U.S. which occurs also in the Ozarks, often near stands of Ozark Witch-hazel. It is best distinguished from the Ozark species by its time of flowering; also its flowers are bright yellow; Sep.–Oct.

The twigs of Eastern Witch-hazel are distilled to produce the antiseptic product of the same name. Some believe that a branch of the shrub can be used to "witch" or "divine" water; i.e., to determine where a well should be dug. When ripe, seed capsules of both species explode, hurling its seeds several feet from the parent plant.

Buck Brush *Andrachne phyllanthoides* (Nutt.) J. Coulter
Spurge Family Euphorbiaceae

The spurge family is primarily a tropical one. In our area it includes principally herbs with a milky sap and inconspicuous flowers.

Buck Brush is the only known native woody spurge of the Ozarks. A short (2–3 ft.) plant with green or reddish stems and small rounded, alternate leaves, it produces tiny (¼ in. across) greenish yellow flowers on stalks ¾ in. long. It is a shrub of glades, bald knobs, and other rocky places of the Ozark Plateau and Ouachita Mountains (apparently absent from OK); May, Jun.

Buck Brush is not to be confused with Coralberry, *Symphoricarpos orbiculatus* Moench, a shrub also sometimes called Buck Brush; see red/orange section.

Tulip-tree
Liriodendron tulipifera

Ozark Witch-hazel
Hamamelis vernalis

Buck Brush
Andrachne phyllanthoides

Red/Orange

Wake Robin *Trillium sessile* L.
Lily Family Liliaceae

Trilliums, noted for their triangular shape, are quite distinctive. The name indicates their 3 broad leaves as well as the 3 sepals and 3 petals of each flower. The half-dozen or so Ozark species can be categorized as either stalked (flower held above leaves by a stalk) or sessile (not stalked). All reddish, purplish, and greenish trilliums known to occur in the Ozarks are described here; all are sessile. Other, stalked, trilliums are featured in the white section.

Also called Toadshade and Sessile Trillium, this 4–12 in. tall plant has maroon-purplish (or greenish) erect (or spreading) petals and green spreading sepals. Leaves are usually, but not always, mottled. This, the most common of the Ozark trilliums, is found in rich woods, especially on slopes and in valleys. Its range in our area includes the Ozark Plateau of MO, AR, and OK (ne. corner only), and IL Ozarks; Apr.–Jun.

Ozark Green Trillium *Trillium viridescens* Nutt.
[*T. viride* Beck var. *viride*]
Lily Family Liliaceae

This variable plant can be separated from Wake Robin (above) by its petals, which are green, becoming reddish toward their bases rather than purplish or maroon. Also, the petals are and more narrow, and taper toward their bases. The plant is not common but occupies rich limestone soils of woods and glades throughout much of the Ozarks except for Crowley's Ridge and the IL Ozarks. It is most abundant in the Boston Mountains of AR, also w. OK and sw. MO; Apr., May.

Green Trillium, *T. viride* Beck, is sometimes confused with Ozark Green Trillium. Its sepals and petals are nearly solid green. Its range includes sw. IL and the adjacent portion of MO; Apr., May.

Prairie Trillium *Trillium recurvatum* Beck
Lily Family Liliaceae

This trillium can best be distinguished from Wake Robin (above) by its recurved sepals (angle downwards); petals are erect, as they are in Wake Robin. It is a somewhat taller plant (6–12 in.), and its mottled leaves taper towards their bases. Also called Purple Trillium, it is found in rich moist woods. Its center of distribution lies ne. of the Ozarks. In our area, it is found in the IL Ozarks, e. and s. MO, and n. and e. AR (absent from OK); Apr.–May.

For a thorough treatment of trilliums of the world—their identification, distribution, and culture—the book *Trilliums* by Fredrick Case and Roberta Case is especially recommended.

Wake Robin
Trillium sessile

Ozark Green Trillium
Trillium viridescens

Prairie Trillium
Trillium recurvatum

Tiger Lily *Lilium tigrinum* L.
Lily Family Liliaceae

This lily, 2–5 ft. tall, has reflexed (curled) tepals and protruding style and stamens. Like those of other *Lilium* species, the tepals are spotted on the inside. Leaves are alternate, with dark bulblets at the nodes. An Asian native, it sometimes escapes from gardens; Jul., Aug.

Michigan Lily, *L. michiganse* Farw., the most common lily of the Midwest, has whorled leaves and lacks the bulblets. It occupies moist sites throughout the Ozarks; Jun., Jul.

Turk's-cap Lily *Lilium superbum* L.
Lily Family Liliaceae

This, the most spectacular lily of e. U.S., has recently been found in the Ozarks. In contrast to Michigan Lily (above), it is taller (to 8 ft.), and has its leaves in whorls of 8–10; also it usually bears 10–40 (vs. 2–3) flowers per plant. Each flower has a centrally located "green star." Note the reflexed petals, which form the "Turk's cap." It has been reported from 3 counties of the Ozark Plateau of AR (Yell, Pope, and Stone) where it occupies moist, partly shaded habitats; Jul., Aug.

Michigan and Turk's-cap Lilies are the only two lilies know to be native to the Ozarks.

Red Iris *Iris fulva* Ker
Iris Family Iridaceae

The flowers of this 1– to 3-ft. plant, which is also called Copper Iris, are orange-red to brownish red (occasionally dull yellow). More common in the swamps of lowlands adjacent to the Ozarks, it occurs in the n. Ouachita Mountains (AR), Crowley's Ridge, and the IL Ozarks.

Yellow Iris, *I. pseudacorus* L., a European species, is very similar but has yellow flowers; May–Jul. It is also a wetland plant.

Blackberry-lily *Belamcanda chinensis* (L.) D.C.
Iris Family Iridaceae

Neither a blackberry nor a lily, this tall (2–3 ft.) plant is related to irises, as indicated by its overlapping, swordlike leaves. Its flowers resemble those of lilies but have only 3 (vs. 6) stamens. In autumn the fruits split open to expose the black seeds, suggesting blackberries. An Asian native, Blackberry-lily has become naturalized through most of the Ozark region. It occurs in rocky soils of open woods, glades, and bluffs; Jul., Aug.; fruits, Aug.–Oct.

Blackberry-lily is easily established in a wildflower garden by sowing seeds in autumn. Cultivars, with flowers of several colors, are available from nursery and seed companies. The roots of the plant are used in traditional Chinese medicine for a large number of ailments.

Tiger Lily
Lilium tigrinum

Turk's-cap Lily
Lilium superbum

Red Iris
Iris fulva

Blackberry-lily
Belamcanda chinensis

Wild Ginger *Asarum canadense* L.
Birthwort Family Aristolochiaceae

In the axil of each pair of cordate leaves is a solitary flower at ground level. Lacking petals, the 3 sepals, each with a long tip, form the conspicuous maroon calyx tube. (The position of the flowers allows the seeds to be easily reached by ants, which disperse them.) Wild Ginger is found in rich woods and along streams throughout much of the Ozarks; Apr., May.

Native Americans used its aromatic roots to make a tea to treat a wide variety of ailments. Modern research has revealed the presence of aristolochic acid, an antitumor compound. The roots have also been used as a substitute for true ginger, which comes from an unrelated tropical plant.

Pawpaw *Asimina triloba* (L.) Dunal
Custard Apple Family Anonaceae

This small deciduous tree typically occurs in large numbers in colonies, or "pawpaw patches." The flowers (1 to 2 in. across), which appear before the foliage, include 3 green sepals that enclose 6 reddish or purplish petals in two cycles. Leaves are nearly a foot long; note that the broadest part is near the apex. Pawpaw is found throughout the Ozarks, where it occupies moist woods, especially along streams and in ravines; Feb.–May; fruits, Jun.–Aug.

The fruits, which resemble small rounded bananas, are edible when ripe (brownish). They are also eaten by a wide variety of animals. Both the leaves and seeds, when crushed, have insecticidal properties.

Spotted Jewelweed *Impatiens capensis* Meerburg [*I. biflora* Walt.]
Touch-Me-Not Family Balsaminaceae

Pendant, brightly colored flowers (1 in. long) are typical of both *Impatiens* species native to the Ozarks. They are common, often together occurring in shaded moist habitats, especially along stream banks.

Growing to a height of 3–5 ft., Spotted Jewelweed has succulent stems and dark green leaves 2–4 in. long. It occurs throughout the Ozarks; Apr.–Oct.

Pale Jewelweed, *I. pallida* Nutt., has pale yellow flowers. It also occurs throughout the Ozarks except for the Ouachitas of AR and Crowley's Ridge of AR; Apr.–Oct.

Both species are also called "Touch-me-not," an allusion to their ripe seedpods, which burst open when touched, an effective mechanism for dispersing its seeds. "Jewelweed" apparently refers to the diffraction of light by drops of water that bead up on its waxy leaves after a rain. The juice of the leaves can be used to counteract the irritation of poison ivy and stinging nettles.

Wild Ginger
Asarum canadense

Pawpaw
Asimina triloba

Spotted Jewelweed
Impatiens capensis

Fire Pink *Silene virginica* L.
Pink Family Caryophyllaceae

Fire Pink is a 1- to 2-ft.-tall plant with opposite, lanceolate leaves. Note the 5 notched petals at right angles to the corolla tube. It is common in open woodlands, especially semishaded banks, throughout but is less common on Crowley's Ridge; Apr.–Jun.

Royal Catchfly, *S. regia* Sims, a spectacular prairie plant, is similar but taller (to 4 ft.) and has unnotched bright red petals; Jun.–Aug.

Red Buckeye *Aesculus pavia* L. [*A. discolor* Pursh]
Horse Chestnut Family Hippocastanaceae

Red Buckeye has compound leaves with 5 leaflets. It may grow to 25 ft. in height. From the flowers are produced smooth, leathery fruits, each with 1–2 seeds (buckeyes). It is typically found in low-lying, moist forests. It occurs in all regions of the Ozarks, but less often on the Ozark Plateau of MO; Apr.–Jun; fruits; Sep.–Oct.

Texas or Ohio Buckeye, *A. glabra* Willd., a medium-sized tree (50–75 ft. tall), has pale yellow flowers followed by spiny fruits; Apr., May; fruits, Sep.–Oct. Its lightweight wood has been used for making cradles, violins, and artificial limbs.

Indian-pink *Spigelia marilandica* L.
Logania Family Loganaceae

Also called Pink-root, this 1- to 2-ft.-tall perennial has 4–7 pairs of sessile, opposite, lanceolate leaves. Above them is a spike of erect tubular flowers each 1–1½ in. long. Note the coral red corolla tube with its 5 yellow lobes. The distinctive wildflower occurs in rich, moist woods and stream banks throughout most of the Ozarks (apparently absent from w. Ozark Plateau of MO); May–Aug.

During the 17th and 18th centuries, the root was used in Western medicine to expel intestinal worms.

Groundnut *Apios americana* Medic.
Pea Family Fabaceae

This twining vine has leaves with 5–7 ovate, sharply pointed leaflets. Its fragrant flowers, clustered in axils of leaves, are maroon or purplish. It occurs in moist to wet soils of thickets and woods throughout the Ozarks; Jun.–Sep.

The protein-rich tubers (swollen underground stems) have been used as a food source by both Native Americans and white settlers (including Pilgrims). They are prepared as potatoes.

Price's Groundnut, *A. priceana* Robinson, which has pale rose-colored flowers, is a rare plant of sw. IL; Sep.

Fire Pink
Silene virginica

Red Buckeye
Aesculus pavia

Indian-pink
Spigelia marilandica

Groundnut
Apios americana

Butterfly-weed *Asclepias tuberosa* L.
Milkweed Family Asclepiadaceae

Like other milkweeds (*Asclepias* species), Butterfly-weed has distinctive, intricate flowers (see pink section for description). Unlike others, it lacks the sticky sap. Also, this is the only milkweed with orange flowers (actually, they vary from yellowish orange through orange to almost red). The numerous opposite leaves are lanceolate. It is a common and conspicuous plant of prairies, glades, and other sunny sites throughout the Ozarks; May–Sep.

The name Pleurisy Root refers to the rhizomes (underground stems) used by Native Americans and early settlers to treat respiratory diseases. However, it is known to contain toxic compounds called cardiac glycosides. It can be transplanted or started from seed in gardens.

Purple Milkweed, *A. purpurascens* L., is a more upright plant with opposite leaves and reddish purple flowers; May–Jul.

Other milkweeds are featured in the white, pink, and green/brown sections.

Red Morning-glory *Ipomoea coccinea* L.
[*Quamoclit coccinea* (L.) Moench]
Morning-glory Family Convolvulaceae

Like other morning-glories, this species is a climbing or trailing vine with trumpetlike flowers. Note the orange corolla tubes and scarlet rims of the trumpets (½ in. across). Leaves may be lobed or unlobed. Introduced from tropical America, it is seen along railroads, in fields, and in other disturbed sites. In our area it is more common on the Ozark Plateau but is also seen in other Ozark regions; Jul.–Oct.

Cypress Vine, *I. quamoclit* L., native to tropical Amer., has been reported as an escape from cultivation. It has similar red flowers but highly dissected leaves; Aug.–Oct. In S. Amer. it is used in folk medicine for sores and snakebites.

Trumpet Creeper *Campsis radicans* (L.) Seem. [*Bignonia radicans* L.]
Trumpet Creeper Family Bigoniaceae

This vine may climb 40–50 ft. up a tree and develop a woody stem 2–3 in. in diameter. Note the compound leaves, each with 7–11 sharply lobed, toothed leaflets. Long (3 in.), trumpetlike flowers are adapted for pollination by hummingbirds. The fruit is a seed capsule that reaches 6 in. in length. Also called Trumpet Vine, it occurs in woods and thickets, along roadsides and railroads throughout the Ozarks; May–Aug. It is sometimes cultivated but can become overly aggressive. "Cow-itch Vine" alludes to the skin rash some people get from contact with the plant.

Cross-vine, *Bignonia capreolata* L., of the same family, is a vine with paired ovate leaflets and coral/cream flowers (see pink section).

Butterfly-weed
Asclepias tuberosa

Red Morning-glory
Ipomoea coccinea

Trumpet Creeper
Campsis radicans

Indian Paint-brush *Castilleja coccinea* (L.) Spreng.
Snapdragon Family Scrophulariaceae

The orange- or scarlet-tipped (rarely yellow or white) 3-lobed bracts are the showy parts of this 1- to 2-ft. annual or biennial. The inconspicuous yellow flowers are all but hidden, except for their protruding pistils. This, the only widespread *Castilleja* species of the Ozarks, is found in moist, grassy open places of the Ozark Plateau and, less often, the Ouchitas; Apr.–Jul.

Like many other members of the snapdragon family, Indian Paint-brush is hemiparasitic: its roots draw nourishment from nearby plants by connections called haustoria. Potentially toxic, its flowers were used by Native Americans to prepare weak solutions as "sweet love charms" and for "female diseases."

Purple Indian Paint-brush, *C. purpurea* G. Don, is a rare perennial known in our area only in Greene Co. (MO) of the Ozark Plateau and e. OK. It has purple or violet-colored bracts; Apr., May.

Wood Betony *Pedicularis canadensis* L.
Snapdragon Family Scrophulariaceae

This low-growing (5–10 in.) perennial has deeply incised leaves, most of which are basal, but a few occur on the flower stalk. Hooded flowers may be whitish, yellow, red, or bicolored red/yellow, as seen here. It is found in dry to wet acidic soils, often along streams, throughout most of the Ozarks; Apr., May.

Another name, "Lousewort," indicates its former use as an insecticide, especially for lice. Native Americans, who considered the roots to be an aphrodisiac, also used a root tea for digestive and heart ailments.

The less common Swamp Lousewort, *P. lanceolata* Michx., is a taller, fall-blooming plant of wet sites. Its lanceolate leaves are dentate but not dissected; Aug.–Oct.

Cardinal Flower *Lobelia cardinalis* L.
Bluebell Family Campanulaceae

Leaves of this spectacular wildflower are alternate, lanceolate, and dentate. The unbranched stem (2–3 ft. tall) has a milky sap. As in other lobelias, each flower (1½ in. long) has 2 lips: upper one of 2 lobes, lower one of 3 lobes. Stamens (5) and style form the central column. Our only red lobelia, it brightens the banks of streams and lakes throughout the Ozarks; Jul.–Oct.

Native Americans used both root and leaf teas to treat a variety of ailments; also as an ingredient in "love potions."

Linnaeus named the genus *Lobelia* in honor of Matthias de l'Obel (1538–1616), a Flemish botanist and herbalist. Other lobelias are featured in the blue/purple section.

Indian Paint-brush
Castilleja coccinea

Wood Betony
Pedicularis canadensis

Cardinal Flower
Lobelia cardinalis

Spice-bush *Lindera benzoin* (L.) Blume
Laurel Family Lauraceae

This highly branched shrub grows 6–12 ft. tall; leaves are 2–5 in. long. Its tiny, yellowish green flowers are arranged in clusters less than ½ in. across. Staminate and pistillate flowers are borne on separate plants. One of the earliest shrubs to flower, it is commonly seen along stream banks and other moist to wet sites throughout most of the Ozarks; Mar.–May; fruits, Aug.–winter.

The shrub in flower is featured in the green/brown section.

Pond Berry, *L. melissifolia* (Walt.) Blume, is a very rare, Federally Endangered shrub reported from Ripley Co. (MO) of the e. Ozark Plateau (and several Delta counties of ne. AR). It is a shorter (to 6 ft.) shrub with thin, drooping leaves that are prominently veined underneath. Steyermark mentions the use by children of its red berries in popguns; Mar., Apr.; fruits, Aug.–winter.

Strawberry Bush *Euonymus americanus* L.
Staff-tree Family Celastraceae

Fruits of this erect or arching shrub are more conspicuous than its small, greenish yellow, 5-petaled flowers. The scarlet, warty fruits break open to expose its orange-red seeds. The opposite leaves are widest near the middle. Also called Brook Euonymus, it occupies moist woods and stream banks. It is relatively common throughout AR, less so in the IL Ozarks and OK; in MO it is known only on Crowley's Ridge; Apr., May; fruits, Nov.–spring.

Two other shrubby *Euonymus* species of the Ozarks that have showy fruits are Wahoo, *E. atropurpureus* Jacq., which has clusters of purplish flowers followed by smooth, rose-colored, 4-lobed fruits, May, Jun.; and Trailing Strawberry-bush, *E. obovatus* Nutt., a prostrate shrub or trailing vine with small greenish flowers that form knobby 3-lobed fruits, Apr.–Jun.

Both Strawberry Bush and Wahoo were used by Native Americans for a variety of medical problems. The bark was used internally as a tonic, diuretic, and laxative; externally, ground to a powder, for dandruff. However, many parts of the plant are probably poisonous if ingested. Beware!

Possum Haw *Ilex decidua* Walt.
Holly Family Aquifoliaceae

Also called Swamp Holly and Deciduous Holly, this species is a shrub or small tree that sometimes becomes 25–30 ft. tall. Note its bluntly tipped leaves. From its clustered white flowers are formed the bright red berries in groups of 5–6. It occurs along streams and fencerows, where it is quite showy in late autumn and winter throughout the Ozarks; Apr., May; fruits, Sep.–spring.

Hollies in flower are featured in the white section.

Spice-bush
Lindera benzoin

Strawberry Bush
Euonymus americanus

Possum Haw
Ilex decidua

Smooth Sumac *Rhus glabra* L.
Cashew Family Anacardiaceae

Sumacs are deciduous shrubs or small trees of roadsides and open fields. Their compound leaves have historically served as a source of tannins for leather, apparently explaining the name "Shoemake" that is sometimes applied. They have clusters of tiny, greenish yellow flowers that produce clusters of hairy, red fruits. All three sumac species described on this page are found throughout most of the Ozarks, where they are among the most colorful plants in autumn.

Smooth Sumac is noted for its smooth twigs and petioles. The large leaves are divided into 11–31 dentate, lanceolate leaflets. Small white or green flowers in a panicle up to 8 in. long are followed by the sharply pointed cluster of hairy fruits seen here; May–Jul; fruits, Jun.–winter.

Winged or Shining Sumac, *R. copallina* L., can be distinguished by its prominent wings that extend along each side of the rachis (the stalk between the leaflets); May–Nov; fruits, Jul.–winter.

Fruits (berries) of sumacs can be used to make a refreshing cold drink that tastes much like lemonade.

Fragrant Sumac *Rhus aromatica* Ait.
Cashew Family Anacardiaceae

Several features distinguish this aromatic sumac from the ones above. Its leaves are, as seen, trifoliate. The small yellow flowers, which appear before the leaves, are in much smaller clusters and on side branches. Fruits (berries) are larger, mature earlier, and are in clusters of a dozen or fewer. It occupies rocky sites including glades, woods, and along railroads (absent from OK); Mar.–Apr.; fruits, May–Jul.

Poison Ivy, *Toxicodendron radicans* (L.) Kuntze, is a common and widespread shrubby vine with variable leaves that may resemble those of Fragrant Sumac. A distinguishing feature is the prominent petiole of each terminal leaflet of Poison Ivy. Also, its berries are white (see white section); May–Jul.; fruits, Aug.–Nov.

Coralberry *Symphoricarpos orbiculatus* Moench
Cashew Family Anacardiaceae

Also called Buck Brush and Indian Currant, this is a low (4–6 ft. tall), branching shrub with ovate, opposite leaves arranged on wiry, arching stems. It is most noticeable in late summer and fall, when its numerous coral-reddish berries form dense clusters around the leaf axils. Flowers are small, yellowish tinged with purple, and trumpet-shaped. Tolerating both sun and shade, it is seen in open woods and rocky banks, where it often forms dense thickets. It occurs throughout the Ozarks; Jul.–Aug; fruits, Aug.–winter.

Smooth Sumac
Rhus glabra

Fragrant Sumac
Rhus aromatica

Coralberry
Symphoricarpos orbiculatus

Common Persimmon *Diospyros virginiana* L.
Ebony Family Ebenaceae

Persimmon is a medium-sized tree (to 50 ft.) with a dark gray bark divided into vertical rectangular blocks. Its leaves, 4–6 in. long, are oval or elliptical, dark green above, pale green beneath. It is a dioecious species: staminate flowers, ³⁄₈ in. long, and pistillate ones, ³⁄₄ in. long, are on separate trees. Both types have pale yellow petals. Fruits, about 1 in. across, that follow pistillate flowers contain 1–3 seeds and ripen as the leaves are shed. Persimmon, along with Sassafras, is a pioneer tree in the process of ecological succession that follows a forest disturbance. It occurs throughout the Ozarks; May, Jun.; fruits, Oct., Nov.

Persimmon fruits are delicious when ripe (but highly astringent before frost). The very hard wood is used in making golf clubs.

Carolina Moonseed *Cocculus carolinus* (L.) DC.
Moonseed Family Menispermaceae

This rambling vine has triangular leaves (as seen here) or cordate ones. Clusters of small whitish flowers are followed by bright red, waxy berries. It is a fairly common plant of woods and thickets throughout the Ozarks; Jul.–Aug.; fruits, Sep.–Dec.

The common name reflects the seeds (one per berry) which are shaped like a three-quarter moon.

Flowering Dogwood *Cornus florida* L.
Dogwood Family Cornaceae

This is the familiar native dogwood often planted as an ornamental. The berries seen here are produced from small greenish flowers surrounded by 4 showy white (or pink) bracts. It is a common understory tree (to 35 ft.) of acidic soils of deciduous forests throughout the Ozarks; Mar.–Jun.; fruits, Aug.–Nov.

Flowering Dogwood, the state tree of MO, is a very attractive ornamental tree at all seasons. In spring the blossomlike bracts make it very conspicuous. In fall both the berries and leaves are, as seen here, bright. At all seasons, especially winter, the pattern of horizontal branches is attractive.

The tree was regarded by Native Americans, and also by some early African and European settlers, as a virtual drugstore. Twigs were chewed as a substitute for a toothbrush; root-bark was used to treat malaria; and the bitter berries for stomach upset. The bark can be used to produce a scarlet dye, as well as black ink. The hard, close-grained wood is useful in the textile industry for shuttles and bobbins.

Dogwoods in flower are featured in the white section.

Common Persimmon
Diospyros virginiana

Carolina Moonseed
Cocculus carolinus

Flowering Dogwood
Cornus florida

Pink

Prairie Onion *Allium stellatum* Ker.
Lily Family Liliaceae

Allium species, which includes such familiar garden plants as onion, garlic, asparagus, and others, includes also several native Ozark plants. The group is characterized by linear leaves, small whitish or pinkish flowers arranged in umbels, and the characteristic strong odor. Most are edible (none are poisonous).

This beautiful onion with rose to pink flowers grows to 2 ft. in height. It occurs on limestone glades and ledges of bluffs of the Ozark Plateau and the IL Ozarks; Jul.–Nov.

Nodding Wild Onion, *A. cernuum* Roth, is remarkably similar to Prairie Onion. The most useful feature for distinguishing it is the crook at the top of the stem caused by the weight of the umbel; Jul.–Sep. Wild Leek or Ramps, *A. tricoccum* Ait., has much wider (1–2 in.) leaves that shrivel just as the flower buds open. Its umbels resemble those of Prairie Onion but are much smaller and are on shorter stalks; Jun., Jul.

Native *Allium* species are well suited to rock gardens.

Rose Pogonia *Pogonia ophioglossoides* (L.) Jussieu
Orchid Family Orchidaceae

Each flower (1 to 3 per stem) of Rose Pogonia has 3 long sepals and 2 shorter, oblong petals. The third, crested, highly fringed petal is distinctive. In addition to the upper leaf seen here, another, larger one is attached at midstem. Also called Snake-mouth, the rare, 4–18 in. plant is known to occur in the Ozarks in these counties: Reynolds (MO), Fulton, Saline, and Jefferson (AR). The usual habitats are wet meadows or bogs, especially those fed by springs; Mar.–Aug.

Nodding Pogonia *Triphora trianthophora* (Sw.) Rydb.
Orchid Family Orchidaceae

This small, delicate orchid, usually less than 6 in. tall, has 2–8 tiny cordate leaves attached to the stem. Another common name, Three Birds Orchid, refers to the 3 (more or less) white to pink nodding flowers (with a superficial resemblance to birds) borne at the top of the plants. Although it has a general range from FL to TX to New Eng., it is not often seen, partly because of its size and habitat requirements, but also because it typically blooms only once every 3 years. In our region it occurs in rich woods near streams, widely scattered throughout the Ozark Plateau (MO, AR) and Ouachita Mountains (AR); Jul.–Sep.

For more information on orchids in general and detailed information on many Ozarkian orchids, see *Wild Orchids of Texas* by Joe Liggio and Ann O. Liggio.

Prairie Onion
Allium stellatum

Rose Pogonia
Pogonia ophioglossoides

Nodding Pogonia
Triphora trianthophora

Pink Knotweed *Polygonum pensylvanicum* L.
Buckwheat Family Polygonaceae

Many *Polygonum* species are called "knotweed" because of a sheath surrounding the nodes. The name "smartweed" is also applied to many of these same plants because of the sharp taste of their (simple) leaves. They have very small, usually pink, petal-like sepals in spikes or spikelike racemes. Many of the nearly two dozen Ozarkian *Polygonum* species are useful as indicators of wetland conditions.

Also called Pennsylvania Smartweed, this erect, branching annual with narrowly lanceolate leaves grows to 5 ft. in height. It is commonly found in open places, especially in rich, moist soil, throughout the Ozarks.

Because of its astringent properties, Native Americans used a leaf tea to stop bleeding and poulticed leaves for piles.

Two rather showy smartweeds of wetlands are also highly variable plants with pink flowers but have wider leaves. Swamp Smartweed, *P. coccineum* Muhl., has flower spikes 2–4 in. long; Jul.–Sep. Water Smartweed, *P. amphibium* L., has flower spikes less than 1 in. long; Jun.–Sep.

Spring-beauty *Claytonia virginica* L.
Purslane Family Portulacaceae

Like other members of the purslane family, Spring-beauty has flowers with 2 sepals, 5 petals, and 5 stamens. Note the pink-veined white petals and linear leaves. As it propagates itself by corms (root swellings), it often occurs in large colonies. It is a common wildflower of rich woods throughout the Ozarks; Feb.–May.

Plants of this species with wider leaves than those seen here have sometimes been confused with Carolina Spring-beauty, *C. caroliniana* Michx., an e. species not known from the Ozarks. The starchy tubers of Spring-beauty were used as a potato substitute by both Native Americans and white settlers.

Large-flowered Fame Flower *Talinum calycinum* Engelm.
Puslane Family Portulacaceae

Fame Flowers, also called Rock Pinks, are small plants with pink flowers. They are able to survive in thin rocky soil by storing water in their tiny succulent leaves.

This species has leaves 2–2½ in. long. Its deep pink flowers, ½ in. across, are supported by wiry stems 6–10 in. long. Each flowers opens for an hour in the afternoon, between 3:00 and 8:30 CST. Though not commonly seen, it occupies rock outcrops throughout our area; May–Aug.

Small-flowered Fame Flower, *T. parviflorum* Nutt., has shorter leaves (less than 2 in.) and smaller flowers (⅜ in. across) on shorter stalks. Also it has fewer stamens (8 or fewer vs. 30 or more); May–Aug.

Pink Knotweed
Polygonum pensylvanicum

Spring-beauty
Claytonia virginica

Large-flowered Fame Flower
Talinum calycinum

Bouncing-bet *Saponaria officinalis* L.
Pink Family Caryophyllaceae

This stout perennial, native to Europe, has numerous white to pink flowers, each of which consists of a corolla tube and 5 slightly notched, reflexed petals. It grows along roadsides and railroad tracks and in other waste places throughout our area; Jun.–Oct.

Rub moistened leaves between your hands and the lather formed explains an alternative name, Soapwort. Bouncing-bet is an old term for a woman who washes clothes; thus both common names refer to the use of the plant as a soap substitute. The same saponins responsible for lathering also can cause severe digestive problems if the leaves are ingested.

Cow Soapwort, *S. vaccaria* L., also an alien plant of waste places, has leaves each with a single prominent central vein and flowers that are generally a deeper pink or rose; May–Sep.

Canadian Columbine *Aquilegia canadensis* L.
Buttercup Family Ranunculaceae

This is the only native columbine of the Ozarks. Each of the graceful inverted flowers has 5 coral spurs with nectar inside, which attracts hummingbirds. Inside are sepals, usually cream-colored (flowers are sometimes all yellow). Leaves are compound with 3 leaflets. It is generally common throughout most of our area but less so on Crowley's Ridge and in OK. Also called Wild Honeysuckle, it occupies limestone bluffs and woods; Apr.–Jul.

Native Americans used crushed seeds as love charms, to treat headaches, and to control lice; roots were chewed to correct various digestive problems.

This species of *Aquilegia* and many of the 20 species of w. N. Amer. are sometimes cultivated.

Leatherflower *Clematis viorna* L.
Buttercup Family Ranunculaceae

Of the half-dozen Ozarkian *Clematis* species of vines, those with thick "petals" (actually sepals) are called leatherflower. Flowers are followed by clusters of plumelike fruits.

Clematis viorna is a climbing vine with cordate leaves. Each of the urn-shaped flowers (1 in. long) is enclosed by 5 dull pink to cream petal-like sepals (petals are absent). Lower leaves are typically divided into 3 rounded leaflets. It is found primarily in limestone woods and on bluffs. In our area it is confined to the Ozark Plateau (especially that of AR) and the IL Ozarks; May–Jul.

Leatherflower, *C. pitcheri* T. & G., has leaves variable in shape but with reticulated (veins form a network) leaves; May–Sep. Swamp Leatherflower, *C. crispa* L., has showier, purplish or lilac flowers; Apr.–Jul.

Virgin's-bower, *C. virginiana* L., is featured in the white section.

Bouncing-bet
Saponaria officinalis

Canadian Columbine
Aquilegia canadensis

Leatherflower
Clematis viorna

Widow's Cross *Sedum pulchellum* Michx.
Orpine Family Crassulaceae

The tiny, white to pink flowers of this plant are arranged in inflorescences with 3–7 branches. The leaves (½ in. across) are narrow and cylindrical. Also called Rock-moss and Lime Stonecrop, it occupies limestone, chert, or sandstone glades and outcrops. Its range includes primarily the Ozark Plateau; it is found less often in the Ouachitas, Crowley's Ridge, and the IL Ozarks; May–Jul.

Yellow Stonecrop, *S. nuttallianum* Raf., is a similar plant of acidic soils. It has yellow flowers; May, Jun. Woodland Stonecrop, *S. ternatum* Michx., has succulent, rounded leaves and white flowers; Apr.–Jun.

Everlasting Pea *Lathyrus latifolius* L.
Pea Family Fabaceae

Members of the pea or legume family are distinguished primarily by the shape and arrangement of their flowers: 5 petals, including 2 lower ones joined to form the keel, 2 lateral ones that form wings, the upper erect one a standard. Fruit is a pealike pod. Leaves are alternate and almost always compound. Legumes are important ecologically because of their ability to convert gaseous nitrogen in the air into nitrogen compounds needed for plant growth.

Everlasting Pea is a sprawling or climbing perennial vine. Leaflets are in single pairs with tendrils attached between each pair. Nonfragrant flowers, which may be pink, white, or purplish, are 2 in. across. A European native, it is seen primarily near old homesites or along roadsides sporadically throughout the Ozarks; May–Sep.

The cultivated Sweet Pea, *L. odoratus* L., is a similar but annual vine with fragrant flowers; May–Sep.

Goat's-rue *Tephrosia virginiana* (L.) Pers.
Pea Family Fabaceae

Also called Hoary Pea because of its silky hairy covering, this unbranched, 1- to 2-ft.-tall plant has distinctive bicolored (yellow/pink flowers). It is common in sunny places with acidic soils such as glades, prairies, and open woods throughout the Ozarks; May–Aug.

Hoary Pea, *T. onobrychoides* Nutt., is a taller (3–4 ft.), more hairy plant. Its flowers change from white to deep red as they mature. In the Ozarks it is confined to the Ouachitas and w. Ozark Plateau of AR and OK; May–Aug.

Tephrosia species are known worldwide for their usefulness as insecticides and fish poisons (after being stunned in the water, fish can be caught by hand). Goat's-rue has also been used for the treatment of tuberculosis and coughs. Modern research reveals that it has potential in cancer treatment.

Widow's Cross
Sedum pulchellum

Everlasting Pea
Lathyrus latifolius

Goat's-rue
Tephrosia virginiana

Spurred Butterfly Pea *Centrosema virginianum* (L.) Benth.
Pea Family Fabaceae

The twining vine, which has trifoliate compound leaves, can climb to 6 ft. Each narrow leaflet is 1–4 in. long. Note the conspicuous (1–1½ in. long) standard (large spreading upper petal) of the flower and the central white spot. In our area it occupies open sites with sandy soils, principally those of the Ouachita Mountains, less often Ozark Plateau of AR (apparently not found in MO, OK, or IL); Jun.–Aug.

Butterfly Pea, *Clitoria mariana* L., also a twining vine, has very similar flowers and compound leaves, but its leaflets are much wider and there are reddish streaks on the flower instead of the white spot; Jun.–Aug.

Sampson's Snakeroot *Psoralea psoralioides* (Walt.) Cory
Pea Family Fabaceae

Also called Scurf-pea, this perennial is a slender plant (1–1½ ft. tall) with spikes of pink to violet flowers. The leaves are divided into 3 narrow leaflets, the middle one with a longer stalk than the lateral ones. It occupies acidic soils formed from granite, chert, and sandstone, including prairies and glades throughout the Ozarks; May–Jul.

French Grass, *P. onobrychis* Nutt., is a less common plant of the n. Ozark Plateau of AR and Crowley's Ridge of MO. It differs by having leaflets that are more rounded; May–Sep.

Sensitive Briar *Schrankia nuttallii* (DC. ex Britt. and Rose) Standl. [*S. uncinata* Willd.]
Mimosa Family Mimosaceae

A sprawling or trailing plant that may reach 4 ft. in length, Sensitive Briar has stems armed with hooked prickles and twice-divided leaves. Like other members of the mimosa family, its leaves quickly fold up when touched. It occupies glades, prairies, roadside banks and other open, sunny sites throughout our area except for the IL Ozarks; May–Sep.

If you walk barefooted through a glade where this plant is abundant, you will appreciate the name Devil's Shoestring given to this plant.

The genus *Schrankia* honors Franz von Paula von Schrank (1747–1835), a German botanist.

Spurred Butterfly Pea
Centrosema virginianum

Sampson's Snakeroot
Psoralea psoralioides

Sensitive Briar
Schrankia nuttallii

Wild Geranium *Geranium maculatum* L.
Geranium Family Geraniaceae

Not to be confused with the potted indoor "geranium" (*Pelargonium* species) native to S. Africa, *Geranium* species are plants of temperate Eurasia and N. Amer. They have dissected leaves and small to medium-sized pink flowers.

Wild Geranium, our largest species (1–2 ft. tall), has hairy 5-lobed leaves and pink to lavender 5-petaled flowers, each an inch across. It is common in rich woods and thickets throughout; Apr.–Jun.

Other *Geranium* species are weedy plants of waste places; their flowers are much smaller. "Cranesbill" refers to the sharply pointed seedpods. Carolina Cranesbill, *G. carolinianum* L., has short-stalked flowers in compact clusters; May–Jul. The alien Dove's-foot Cranesbill, *G. molle* L., is a low plant with bent seedpod stalks and its rounded leaves divided into short, blunt lobes; Jun.–Aug.

Rose-pink *Sabatia angularis* (L.) Pursh
Gentian Family Gentianaceae

Also called Rose Gentian, this highly branched, 1- to 3-ft.-tall biennial has thick, 4-angled stems. The opposite leaves are large and broadly lanceolate near the bottom of the plant, becoming smaller upward. The fragrant flowers, about 1 in. across, vary from the pink seen here to almost white; the greenish star in the middle of each is characteristic of *Sabatia* flowers. Rose-pink is fairly common in acid soils of glades and open woods throughout most portions of the Ozarks; Jun.–Aug.

Prairie Rose Gentian, *S. campestris* Nutt., is a shorter (to 9 in.) plant; its flowers have longer calyx lobes than those of Rose-pink. It occurs in fields and prairies; Jul.–Sep.

Downy Phlox *Phlox pilosa* L.
Phlox Family Polemoniaceae

Phloxes are upright or trailing leafy perennials with opposite leaves and small but showy flowers that are usually pink or blue. Each flower includes 5 petals united into a corolla tube from which 5 separate petals extend at right angles.

The very narrow leaves of this 10- to 20-in.-tall perennial help to distinguish it from other phloxes. Note also the wedge-shaped petals, each about ¾ in. wide. "Downy" refers to the softly pubescent corolla tubes. It occupies rocky woods, prairies, and other open sites throughout the Ozarks; Apr.–Jul.

Garden Phlox, *P. paniculata* L., is a taller (to 5 ft.) plant with much larger, wider, and more veiny leaves; Jul.–Oct. Smooth Phlox, *P. glaberrima* L., is similar to Garden Phlox but, in addition to flowering earlier, has leaves that are narrower and more pointed; May, Jun.

Wild Geranium
Geranium maculatum

Rose-pink
Sabatia angularis

Downy Phlox
Phlox pilosa

Slender Gerardia *Agalinis tenuifolia* (Vahl) Raf. [*Gerardia tenuifolia* Vahl]
Snapdragon Family Scrophulariaceae

Gerardias have pink or purplish bell-like flowers arranged in pairs at the axils of opposite linear leaves. Slender Gerardia is a plant that grows to 2 ft. tall; its leaves are very narrow ("*tenuifolia*") and only 1¼ in. long. Flowers are ¾ in. across. It is found in wet prairies and wetlands or on drier sites throughout most of the Ozarks (apparently absent in OK); Aug.–Oct.

Gattinger's Gerardia, *G. gattingeri* Small, a plant of sterile acid woodlands, has yellow-green leaves and smaller flowers (½ in. across); Aug.–Oct.

Similar plants with yellow flowers are now assigned to the genus *Aureolaria.*

Halbert-leaved Rose Mallow *Hibiscus laevis* Allioni [*H. militaris* Cav.]
Mallow Family Malvaceae

Members of the mallow family, which include cotton, okra, and hollyhock, are recognized by numerous stamens that surround and are attached to the style.

This is a tall (3–5 ft.) plant with large (2–4 in. across), showy flowers. Note the deep pink flower center and also the long lanceolate leaves, each with 2 basal lobes. It occupies wetlands throughout the Ozarks; Jul.–Oct.

Hairy-fruited Rose Mallow, *H. lasiocarpos* Cav., also widespread, differs primarily by its leaves, which are cordate, unlobed, and with toothed margins; Jul.–Oct. Flower-of-the-hour, *H. trionum* L., is a smaller (1–2 ft.), weedy plant from Europe that grows in fields and waste places. Its leaves are deeply dissected with 3 narrow leaflets with jagged edges; flowers are cream with a dark area inside; Jun.–Sep.

Maryland Meadow Beauty *Rhexia mariana* L.
Melastoma Family Melastomataceae

Note the 4 asymmetrical, pale pink petals, curved anthers, and narrow, lanceolate leaves. The 1- to 2-ft. plant occupies moist to wet open sites, usually on sandy soils. More common in the Ouchitas (AR, OK), its range extends n. to include also the Ozark Plateau (MO), Crowley's Ridge, and the IL Ozarks; Jun.–Oct.

Virginia Meadow Beauty, *R. virginica* L., is a smoother plant with wider leaves and petals that are a deep rose; Jun.–Oct.

The tart leaves of meadow beauty can be added to salads or boiled as a potherb. The tubers are eaten raw.

Slender Gerardia
Agalinis tenuifolia

Halbert-leaved Rose Mallow
Hibiscus laevis

Maryland Meadow Beauty
Rhexia mariana

Common Milkweed *Asclepias syriaca* L.
Milkweed Family Asclepiadaceae

In addition to their latexlike sap, milkweeds have small, intricate flowers, arranged in umbels. A typical flower has 5 reflexed petals above which are 5 erect hoods that form the corona (crown). Fruits are pods that enclose numerous plumed seeds.

Common Milkweed is a stout, 3- to 5-ft.-tall plant with large (6–8 in. long) ovate leaves. Flowers vary from white to dark pink. The pods are warty and pointed. It occupies open, disturbed sites in counties along the n. edge of the Ozark Plateau (MO), in the IL Ozarks, and sporadically on Crowley's Ridge; May–Aug.

The young shoots and tender pods can be cooked as vegetables.

Blunt-leaved Milkweed, *A. amplexicaulis* Sm., has wavy leaves and flowers that are green/magenta; Apr.–Jul.

Four-leaved Milkweed *Asclepias quadrifolia* Jacq.
Milkweed Family Asclepiadaceae

The name refers to leaves at midstem that are in whorls of 4; those above are paired. Flowers of this 1- to 1½-ft. milkweed vary from white to pink or lavender. It occurs in rocky, open woods of ridges and upland slopes of the Ouachita Mountains (AR), the Ozark Plateau (MO, AR, OK), less often on Crowley's Ridge and the IL Ozarks; May–Jul.

The less common Swamp Milkweed, *A. incarnata* L., is a taller plant (2–4 ft.) with dull pink flowers and narrower, opposite leaves; also its habitat is quite different; May–Jul.

False Dragonhead *Physotegia virginiana* (L.) Benth.
Mint Family Lamiaceae

An overview of the mint family appears in the blue/purple section. This wiry upright perennial can grow to 4 ft. but is usually less than 3 ft. Leaves 3–5 in. long are lanceolate with dentate margins. Flowers (1 in. long) vary in color from light pink to deep rose; each has a spotted lower lip. It is common along riverbanks and other moist sites throughout most of the Ozarks (less common in the Ouchitas); May–Sep.

Wood-sage *Teucrium canadense* L.
Mint Family Lamiaceae

This variable perennial, which may grow from 1 to 3 ft. tall, has dentate lanceolate leaves that vary from broad to narrow. Flowers may be pink, lavender, or cream-colored. Also called Germander and Wild Basil, it is a plant of prairies, meadows, and thickets throughout the Ozarks; Jun.–Sep.

A leaf tea of Wood-sage has traditionally been used as a diuretic, to induce menstruation and sweating, and externally as a gargle and an antiseptic.

Common Milkweed
Asclepias syriaca

Four-leaved Milkweed
Asclepias quadrifolia

False Dragonhead
Physotegia virginiana

Wood-sage
Teucrium canadense

Rose Verbena *Verbena canadensis* (L.) Britt.
Verbena Family Verbenaceae

Having weak stems, this plant, also called Eastern Verbena, usually sprawls on the ground. Leaves are simple but deeply lobed and toothed. Each flower has 5 petals that spread from the corolla tube, suggesting a phlox. It grows in limestone glades and prairies and along roadsides throughout the Ozarks; Mar.–Nov.

These verbenas have small pinkish or purplish flowers in erect spikes; their opposite leaves are sessile. Hoary Vervain, *V. stricta* Vent., has dentate ovate leaves; May–Sep. Narrow-leafed Vervain, *V. simplex* Lehm., has narrowly lanceolate leaves that taper toward their bases; May–Sep. European Vervain, *V. officinalis* L., differs by its lower leaves, which are dissected into sharply pointed lobes; May–Oct. All inhabit open, usually disturbed, sites.

Rosy Milkwort *Polygala cruciata* L.
Milkwort Family Polygalaceae

Milkworts have clusters of inconspicuous flowers resembling those of clovers. Most species have pinkish flowers, but those of some are white or orange.

Also called Cross-leaved Milkwort, this uncommon species has narrow leaves in whorls of 4. The unusual flower spikes are likened by Carl G. Hunter in his *Wildflowers of Arkansas* to "small paper lanterns." Less than 1 ft. tall, Rosy Milkwort occurs sporadically, primarily in sandy soils, in the Ouachitas and Ozark Plateau of AR; Jul.–Oct.

Field Milkwort, *P. sanguinea* L., is a more common species with smaller white to rose-colored flower heads and alternate leaves; May–Oct.

Hollow Joe-Pye-weed *Eupatorium fistulosum* Barratt
Aster Family Asteraceae

Several tall *Eupatorium* species are called Joe-Pye-weed. (Joe Pye was a 19th-century New Englander who promoted the medicinal uses of plants, which he had learned from Native Americans.) They are 2- to 7-ft.-tall, thick-stemmed plants with pink or purple flower heads in dense clusters. Leaves are toothed and whorled. They inhabit wet sites.

Hollow Joe-Pye-weed has rounded flower clusters and hollow purple stems. It is found in the Ouachitas, s. Ozark Plateau (AR and OK), less often on Crowley's Ridge and the IL Ozarks; Jul.–Sep.

Sweet Joe-Pye-weed, *E. purpureum* L., has solid green stems and more flat-topped flower clusters; Jul.–Sep.

Joe-Pye-weeds were used by Native Americans to treat a variety of urinary tract ailments, especially kidney stones. Recent studies in Germany indicate their potential for stimulating the immune system.

Other *Eupatorium* species are featured in the white section.

Rose Verbena
Verbena canadensis

Rosy Milkwort
Polygala cruciata

Hollow Joe-Pye-weed
Eupatorium fistulosum

Prairie Rose *Rosa setigera* Michx.
Rose Family Rosaceae

The rose family is a large, temperate one that includes herbs, shrubs, and trees. Flowers typically have 5 green sepals, 5 rounded petals of various colors, and numerous stamens and pistils. Their compound alternate leaves bear prominent stipules. Roses (*Rosa* species) native to the Ozarks are shrubs with similar flowers (2–3 in. across and with 5 pink petals); thus they are identified to species primarily on the basis of their vegetative features.

Also called Climbing Rose, this is a climbing or sprawling plant distinguished by its 3 leaflets per leaf and solitary, reflexed hooks on the stem. It inhabits open woods and thickets throughout the Ozarks; May–Jul.

These are also native roses. Swamp Rose, *R. palustris* Marsh., has 5 or 7 leaflets and paired, reflexed spines; May–Jul. Pasture Rose, *R. carolina* L., has 5–7 leaflets but slender, straight spines; May, Jun.

Multiflora Rose, *R. multiflora* Thunb., has smaller white flowers in clusters and toothed stipules. An Asian garden escape, it has become a troublesome pest; May, Jun.

A tea made from rose hips (fruits) is high in vitamin C.

Steeple-bush *Spirea tomentosa* L.
Rose Family Rosaceae

Although it has the initial appearance of a herbaceous wildflower, this plant has woody stems that support steeple-shaped clusters of tiny rose-colored (or white) flowers. Alternate leaves are woody and brownish underneath. Also called Hardtack, the 2- to 4-ft.-tall plant requires moist acidic soils. In the Ozarks, it occurs only sporadically on Crowley's Ridge and in central AR near the Arkansas River; Jun.–Aug.

This native plant is sometimes cultivated as a garden ornamental. Leaves of this and other species of *Spirea* were used by Native Americans for the treatment of dysentery, diarrhea, and morning sickness.

Trumpet Honeysuckle *Lonicera sempervirens* Ait.
Honeysuckle Family Caprifoliaceae

The whorls of long (1–1½ in.) slender trumpets, coral with yellow lobes, are distinctive. Note also the paired leaves, said to be perfoliate, "pierced" by the stem. Also called Coral Honeysuckle, this smooth, twining vine is native to e. U.S. but often escapes from cultivation in our area and is usually seen near houses; Apr.–Jul.

The genus *Lonicera* commemorates Adam Lonitzer, a 16th-century German herbalist. Other honeysuckles are featured in the white section.

Prairie Rose
Rosa setigera

Steeple-bush
Spirea tomentosa

Trumpet Honeysuckle
Lonicera sempervirens

Eastern Redbud *Cercis canadensis* L.
Caesalpinia Family Caesalpiniaceae

Before the leaves of this small tree appear, red buds open to form rose-lavender (rarely white) pealike flowers. Flattened seedpods that follow, 2–3 in. long, persist throughout the winter. Leaves are simple, cordate, and 3–5 in. long. Redbud is a common understory tree along the edges of woods and glades and on rocky bluffs throughout the Ozarks; Mar.–May; fruits, May–spring.

A striking floral display occurs in the Ozarks, usually in late April, when the flowering time of Redbud overlaps that of Flowering Dogwood. The flowers give color and a pleasant tartness to salads. The tender young seedpods can be sautéed in butter. Eastern Redbud is the state tree of Oklahoma.

Cross-vine *Bignonia capreolata* L.
Bignonia Family Bignoniaceae

This woody vine, which often climbs to the top of tall trees, has compound evergreen leaves, each with 2 ovate lateral leaflets and a terminal tendril. Note the large (3 in. long), orange to coral, trumpetlike flowers with yellow throats. Its fruits are long (4–6 in.), flat capsules with small, winged seeds. Cross-vine occupies moist soils of fence rows, thickets, and fields throughout the Ozarks with the exception of the n. Ozark Plateau (MO); Apr.–Jun.; fruits, Sep.–Dec.

The common name refers to the X visible in a cross-section of its stem.

Mountain Azalea *Rhododendron roseum* (Lois.) Rehd. [*R. prinophyllum* Millais]
Heath Family Ericaceae

Heaths are shrubs or small trees with urn-shaped flowers and simple, alternate leaves. They are generally restricted to acidic soils. In the Ozarks are several *Rhododendron* species, all deciduous shrubs, and with showy white or pink flowers that generally appear before the leaves. Leaves are at or near the tips of the twigs.

Also called Wild Azalea, Election Pink, and (locally) Honeysuckle, this species, which grows to 7 ft. tall, has pink or rose (rarely white) flowers with long tubular corollas and conspicuous stamens. Leaves, 3–4 in. long, are leathery and hairy underneath. It occupies primarily n.-facing wooded slopes and ravines along streams throughout the Ozarks (except for Crowley's Ridge); Apr., May.

Hoary Azalea, *R. canescens* (Michx.) Sweet, of the Ouachitas (AR, OK), has flowers with pale pink tubes and whitish lobes; Mar., Apr. Texas Azalea, *R. viscosum* (L.) Torr., of the Ozark Plateau (AR, OK), has white flowers that appear after the leaves; Apr., May.

Eastern Redbud
Cercis canadensis

Cross-vine
Bignonia capreolata

Mountain Azalea
Rhododendron roseum

Blue/Purple

Asiatic Dayflower *Commelina communis* L.
Spiderwort Family Commelinaceae

Unlike spiderworts (below), which have 3 petals of equal size, dayflowers have 2 prominent ones and a third odd, smaller one. The latter supports the curved stamens. Each flower lasts only a day. Leaves of dayflowers can be cooked and eaten as a vegetable.

In this alien species, the odd, whitish petal is transparent and beneath the 2 more prominent blue ones. The common sprawling, weedy annual is frequent in waste places and gardens throughout our area; May–Oct.

Slender Dayflower, *C. erecta* L., is a native perennial with similar flowers but an upright growth habit (to 3 ft.); May–Oct.

These dayflowers, also common in our region, have all 3 petals that are blue. Virginia Dayflower, *C. virginiana* L., is a tall (to 3 ft.), erect perennial with large leaves; May–Sep. Spreading Dayflower, *C. diffusa* Burm.f., is a smaller, spreading annual with smaller flowers; Jun.–Oct.

Ohio Spiderwort *Tradescantia ohiensis* Raf.
Spiderwort Family Commelinaceae

Spiderworts are perennials with terminal clusters of flowers (1 in. across) with 3 identical blue, purple, rose, or whitish petals. Frequent hybridization often makes identification difficult.

Ohio Spiderwort is a tall (to 3 ft.), upright, branching plant with smooth flower stalks and buds. It is generally the most common spiderwort of our area (although absent from Crowley's Ridge and OK), where it is seen along roads and railroads, and in open fields; May–Jul.

These spiderworts are also tall plants with, usually, deep blue or purple flowers. Virginia Spiderwort, *T. virginiana* L., differs from Ohio Spiderwort by having hairy flower stalks and buds; Apr.–Jun. Zigzag Spiderwort, *T. subaspera* Ker.-Gawl., has zigzag stems; Jun.–Sep. Both are less common than Ohio Spiderwort.

Wild Crocus *Tradescantia longipes* Anderson & Woodson
Spiderwort Family Commelinaceae

This rare spiderwort is generally easily recognized by its dwarf size; its fragrant flowers, which vary in color from rose to blue to purple, lavender, or even white, are borne barely above ground level. It occurs in woodlands with acidic soils, especially in the St. Francois Mountains of e. MO; it has also been reported from several counties of AR; Apr., May.

Two other species, Ozark and Woodland Spiderworts, which sometimes also have blue/purple flowers, are described in the white section.

Spiderworts can be used as indicators of nuclear radiation; filament hairs change from blue to pink in response to low-level dosages. They are also cultivated as garden ornamentals.

Asiatic Dayflower
Commelina communis

Ohio Spiderwort
Tradescantia ohiensis

Wild Crocus
Tradescantia longipes

Pickerelweed *Pontederia cordata* L.
Pickerelweed Family Pontederiaceae

Pickerelweed, with its cordate leaves on long stalks and dense spikes of purple flowers, is quite showy. It grows rooted on the bottom of shallow ponds and along the edge of lakes and streams, a habitat where pickerel (fish) swim and lay their eggs; thus, Pickerelweed. It is not common but is found sporadically throughout our area; Jun.–Nov.

Found on separate plants are 3 kinds of flowers that differ in the lengths of their styles and stamens. This arrangement, known as "tristyly," promotes cross-pollination. The fruits can be eaten raw when ripe.

The genus *Pontederia* was named in honor of Guilio Pontedera (1688–1756), an Italian botanist.

Wild Hyacinth *Camassia scilloides* (Raf.) Cory
Lily Family Liliaceae

This attractive plant has long (1 ft.) linear leaves that are separate from stalks that bear panicles of pale blue flowers. Each flower consists of 6 pointed tepals. Wild Hyacinth occurs on limestone soils of prairies, glades, and open woods throughout most of our area (except, apparently, Crowley's Ridge); Apr., May.

A much less common species, *C. angusta* (Englem. & Gray) Blankenship, also called Wild Hyacinth, has more numerous (50–100) lavender to light purple flowers, and blooms later; May, Jun.

Dwarf Crested Iris *Iris cristata* Ait.
Iris Family Iridaceae

Irises are named for the Greek goddess of the rainbow, reflecting the wide color variations observed in these plants. Iris flowers appear to have 9 petals; in reality, the outer 3 parts are sepals (called falls), the next 3 are erect petals (standards), and the inner 3 are colorful branches of the stigma. The prominent markings on the falls help direct insects or birds to the nectaries, where they make contact with the stamens and stigmas.

This is a small, low-growing iris; leaves are only 6–9 in. long, and flowers are only 2½ in. across. Note the "bearded" sepals. Plants with white flowers are designated as var. *alba*. It often occurs as extensive colonies in moist woods along bluffs and stream banks. It is found throughout most of our area (except for Crowley's Ridge); Apr., May.

Native Americans made from the root an ointment that was used to treat skin cancer; a root tea for hepatitis.

Dwarf Iris, *I. verna* L., resembles the above but has more slender leaves and its sepals are not bearded. A disjunct species from e. U.S., it occurs only sporadically in the Ouachitas, usually in upland, disturbed sites; Mar.–May.

Pickerelweed
Pontederia cordata

Wild Hyacinth
Camassia scilloides

Dwarf Crested Iris
Iris cristata

Short-Stemmed Iris *Iris brevicaulis* Raf.
Iris Family Iridaceae

About 2 ft. in height, this iris tends to have reclining stems. Flower stalks are attached to the side of the main stem, unlike other iris flowers, which are terminal. Falls are conspicuously marked but are not crested. It is a relatively rare plant of wetlands, known only from a few counties along the w. and e. edges of the Ozark Plateau of MO and AR; Apr., May.

Southern Blue Flag *Iris virginica* L.
Iris Family Iridaceae

This tall (3 ft.) iris bears flowers about 3 in. across. Color of the slender petals and sepals varies from blue or purple to white with blue markings as seen here. It occurs in wetlands and edges of lakes throughout most of our area (absent in OK); May–Jul.

Yellow Iris, *I. pseudacorus* L., a European native, sometimes escapes from cultivation and occurs in much the same habitats. It is a similar plant except its flowers are bright yellow; May–Jul.

Although irises were often used medicinally by Native Americans, they are considered poisonous.

Pointed Blue-eyed Grass *Sisyrinchium angustifolium* P. Mill
Iris Family Iridaceae

Sisyrinchium species are stiff, clump-forming plants with grasslike leaves. Each of the 6 tepals is tipped with a small, sharp bristle. Flower colors vary from deep blue to violet to white. They generally occur in open, moist sites.

This species has long-stalked light blue flowers atop flat, narrow (¼ in.), branching stems, 6–12 in. tall. It occupies moist sites, especially along streams, throughout our area; May–Jul.

Blue-eyed Grass, *S. atlanticum* Bickn., also has blue flowers on branching stems but is a more slender plant with much shorter and more narrow leaves; May–Jul.

The following are unbranched plants with white or pale blue flowers. Prairie Blue-eyed Grass, *S. campestre* Bickn., has flowers that emerge from long green bracts that surround the stem; Apr.–Jun. White Blue-eyed Grass, *S. albidum* Raf., occurs also throughout our area but is less common. Its green bracts enclose 2 (vs. 1 for *S. campestre*) spathes that surround the flower; May, Jun.

Short-stemmed Iris
Iris brevicaulis

Southern Blue Flag
Iris virginica

Pointed Blue-eyed Grass
Sisyrinchium angustifolium

Showy Lady's-slipper *Cypripedium reginae* Walt.
Orchid Family Orchidaceae

The typical orchid flower has 3 sepals and 3 petals, 2 lateral ones with the third modified to form a sac, lip, or spur.

One of the largest lady's-slippers, this 1- to 3-ft.-tall orchid typically bears 2 flowers, each with 3 white sepals; the inflated lip is also white but marked with broad bands of pink or rose purple. It is found only rarely in widely scattered, usually remote locales of the Ozark Plateau of MO and AR. It requires a moist to wet habitat; May, Jun.

Showy Orchis *Galearis spectabilis* (L.) Raf. [*Orchis spectabilis* L.]
Orchid Family Orchidaceae

This distinctive orchid has 3–12 small (½ in. wide) purple-and-white flowers on a 6–10 in. stalk. Note the pair of basal leaves. Not as common as formerly, it occurs infrequently in oak-hickory forests throughout most of the Ozark Plateau (MO, AR), Crowley's Ridge, and the IL Ozarks; Apr.–Jun.

Lily-leaved Twayblade *Liparis lilifolia* (L.) Richard
Orchid Family Orchidaceae

The 4- to 10-in.-long raceme of this plant includes 12–24 delicate purplish or mauve flowers. Note the pair of glossy basal leaves that account for its common name. It occurs infrequently in acidic soils of moist woods or along streams of the Ozark Plateau of MO and AR; less commonly on Crowley's Ridge and the IL Ozarks; May, Jun.

The rare Yellow Twayblade, *L. loeselii* (L.) Richard, is a smaller plant with smaller, yellow-green flowers. It grows in swampy meadows but has been reported from only 2 Ozark counties: Shannon (MO), Ozark Plateau; and Garland (AR), Ouachita Mountains; May, Jun.

Purple Fringeless Orchid *Platanthera peramoena* Gray
[*Habenaria peramoena* Gray]
Orchid Family Orchidaceae

This tall plant (2–3 ft.) includes a stalk with 12–20 showy rose to purple flowers. The odd petal of each flower is delicately fringed. It occurs in moist woods and lake margins of the St. Francois Mountains section of the Ozark Plateau (MO), central AR near the Arkansas River, Crowley's Ridge, and the IL Ozarks; Jun.–Sep.

The name "fringeless" distinguishes it from Purple Fringed Orchid, *P. psycodes* (L.) Lindley, of Crowley's Ridge which is much more deeply fringed. Ragged-fringed Orchid, *P. lacera* (Michx.) Don, a rare plant of Crowley's Ridge bogs, has deeply cut greenish white flowers; May–Sep.

Other *Platanthera* species are featured in the yellow and green/brown sections.

Showy Lady's-slipper
Cypripedium reginae

Showy Orchis
Galearis spectabilis

Lily-leaved Twayblade
Liparis lilifolia

Purple Fringeless Orchid
Platanthera peramoena

Dwarf Larkspur *Delphinium tricorne* Michx.
Buttercup Family Ranunculaceae

Larkspurs are so named because one of the petal-like sepals forms a "spur"; petals are absent. Leaves are dissected somewhat like those of buttercups (*Ranunculus* species).

This 1- to 2-ft.-tall plant produces flowers that vary from purple to pink to white. It is found in rich woods and stream banks throughout the Ozarks; Apr.–Sep.

Tall Larkspur, *D. exaltum* Ait., is a taller plant (2–6 ft.) known only in the Ozarks from Shannon and Howell Cos. of southcentral MO; possibly present in extreme northcentral AR. It has blue or white flowers and similar but larger leaves; Jul., Aug.

Trelease's Larkspur *Delphinium treleasei* Bush
Buttercup Family Ranunculaceae

This markedly handsome plant (to 3 ft. tall) bears its deep blue/purple flowers (1 in. long) in large racemes. Leaves are similar to those of Dwarf Larkspur (above) but divided into narrower segments. Endemic to the Ozarks, it occupies limestone soils of glades and bald knobs, principally of the Ozark Plateau of MO and AR; May, Jun.

Carolina Larkspur, *D. carolinianum* Walt., differs primarily by its flowers (blue to whitish), which are attached to the stem by much shorter pedicels; May, Jun. Ozark Larkspur, *D. newtonianum* Walt., is a rare endemic known only from a few counties of AR: Newton, Searcy, Johnson, Pope, and Van Buren (Ozark Plateau); and Pike, Polk, and Montgomery (Ouchitas). About 2 ft. tall, the plants bear one or several flowers on each long pedicel. Leaf segments are wider than those of either *D. treleasei* or *D. carolinianum;* Jun., Jul.

Blue False Indigo *Baptisia australis* (L.) R. Br.
Pea Family Fabaceae

Plants of the pea family are generally recognized by their compound leaves and flowers that suggest those of sweet peas, followed by beanlike pods. Included are trees, shrubs, vines, and herbs.

Baptisia species are typical of the Fabaceae: pealike flowers and cloverlike leaves. Flowers of this species are large (1 in. across), dark blue or violet, in erect racemes. Their rounded leaflets are wider toward their tips. The smooth conspicuous perennial, 3–4 ft. tall, occupies limestone glades and prairies of the Ozark Plateau; May, Jun.

Blue False Indigo can be grown from seed. Children sometimes use the ripe podlike fruits as rattles. Native Americans poulticed its roots to reduce swellings; a root tea was used as a purgative and emetic.

This is the only blue-flowered *Baptisia* species of the Ozarks.

Dwarf Larkspur
Delphinium tricorne

Trelease's Larkspur
Delphinium treleasei

Blue False Indigo
Baptisia australis

Kudzu *Pueraria lobata* (Willd.) Ohwi
Pea Family Fabaceae

This noxious weedy vine is a native of Japan. Each leaf has 3 wide 3-lobed leaflets. Racemes of sweet-smelling violet flowers, often hidden behind the leaves, are only 3–4 in. long. It is common in the se. U.S. and in the Ozarks. It was often planted for erosion control and as a source of starch and fiber; Aug.–Sep.

Kudzu, which smothers crops and native vegetation in the U.S., is widely used in traditional Chinese medicine. The genus was named for the Swiss botanist M. W. Puerari (1765–1845).

Indigo Bush *Amorpha fruticosa* L.
Pea Family Fabaceae

Indigo Bush, also called False Indigo, is an erect shrub with deciduous leaves that are fragrant when crushed. The numerous (13–25) leaflets have small dots underneath. Flower racemes are to 6 in. long. The purplish blue flowers are unusual in that each has only one petal. Fruits are less than ½ in. long and warty. Growing to 15 ft. tall, it forms thickets along stream banks throughout the Ozarks; Apr.–Jun.

Three other *Amorpha* species have solitary racemes. Lead Plant, *A. canescens* Pursh, is a shorter (less than 3 ft.), more branched shrub with smaller, hairy, grayish green leaves. It occupies prairies and glades; Apr.–Aug. Shining Indigo Bush, *A. nitens* Boynton, is a smooth (hairless) plant with glossy, oval leaflets; Apr.–Aug. Ouachita Leadplant, *A. ouachitensis* Wilbur., resembles the latter but has acute (pointed vs. rounded) calyx lobes; Apr.–Aug.

American Wisteria *Wisteria frutescens* (L.) Poir [*W. macrostachya* Nutt.]
Pea Family Fabaceae

This twining woody vine often climbs high into trees. Its racemes of fragrant purplish flowers are 4–6 in. long. Seedpods that follow are smooth (hairless), reddish brown with prominent constrictions between the large seeds inside. It occupies moist to wet sites along streams and swamps in the Ozarks of AR (less common in MO and the IL Ozarks); Apr.–Jun.

Two Asian wisterias sometimes escape from cultivation and are seen in the Ozarks of AR. Both have seedpods with a velvetlike surface due to many small hairs. Japanese Wisteria, *W. floribunda* (Willd.) DC., has racemes of flowers 8–15 in. long and leaves with 13–19 leaflets; Apr., May. Chinese Wisteria, *W. sinensis* Sweet, has racemes 6–8 in. long and leaves with 7–13 leaflets; Apr., May.

Although very attractive when in flower, Asian wisterias are aggressive plants that smother native vegetation. Wisteria seeds are highly poisonous (2–3 seeds if ingested may kill a child!); other parts are potentially toxic.

Kudzu
Pueraria lobata

Indigo Bush
Amorpha fruticosa

American Wisteria
Wisteria frutescens

Common Blue Violet *Viola sororia* Willd. [*V. papilionacea* Pursh]
Violet Family Violaceae

Violets, with their small, 5-petaled flowers (2 upper, 2 lateral, and 1 terminal) and simple leaves, are among the most recognizable groups of wildflowers. Identification to species, however, is often difficult because of variations within and hybridization between species. Features to consider in identification include leaf shape, flower color and markings, and whether the violet is "stemmed" (leaves, flowers on same stem) or "stemless" (on separate stems). In addition to the dozen species native to the Ozarks, there are also several European ones that you might encounter.

This smooth or downy, stemless violet is highly variable. A "typical" plant of the variable species, such as seen here, is 4–9 in. tall and has cordate leaves; both lateral petals are bearded, and the terminal petal is slightly longer and unbearded. It occurs in woods and thickets throughout the Ozarks (probably our most common violet); Mar.–Jun.

Missouri Violet, once considered a separate species, is now considered by violet specialists Landon McKinney and Kurt Blum to be the variety *missouriensis* of *V. sororia.* Its leaves are nearly triangular and its petals a pale lavender or lilac; Mar.–May.

Plains Violet *Viola palmata* L. (*V. viarum* Pollard)
Violet Family Violaceae

The flowers of this stemless violet are very similar to those of Common Blue Violet (above). Note the 3-lobed leaves that are longer than the flower stalks. It occupies moist places, commonly seen along roadsides throughout the Ozarks (but probably absent in OK and the IL Ozarks); Apr., May.

Arrow-leafed Violet, *V. sagittata* Ait., has flower stalks as long as the arrow-shaped (each of its 2 basal lobes has 2–3 sharp tips) leaves; Apr.–Jun.

Birdfoot Violet Viola pedata L.
Violet Family Violaceae

Also a highly variable, stemless violet, this low-growing (6 in.) but spreading plant has deeply dissected leaves that resemble a bird's foot. Its flowers are large (1–1¼ in. across), beardless; in one variety they are bicolored (purple/lavender), and in another, white. Also called Pansy Violet, it occurs in rocky, usually acidic, soils throughout the Ozarks; Apr.–Jun.

The leaves of Birdfoot Violet were used by Native Americans to prepare an expectorant and for certain lung complaints; during the 19th century it was used in a similar way by European physicians.

Steyermark, in his *Flora of Missouri,* describes a game called "Hens and Roosters" played by Ozark children using flowers of the solid purple type (hens) and bicolored ones (roosters). Other violets are featured in the white and yellow sections.

Common Blue Violet
Viola sororia

Plains Violet
Viola palmata

Birdfoot Violet
Viola pedata

Passion Flower *Passiflora incarnata* L.
Passion Flower Family Passifloraceae

A member of a principally tropical family, this sprawling or climbing vine is one of the few species of the family in temperate North America. The flowers (to 2 in. across) are not easily described but are so distinctive as to be easily recognized; the corona (an outgrowth between the stamens and petals) is divided into numerous fringelike purple/white segments. Passion Flower occurs in fencerows and other sunny, disturbed places throughout the Ozarks (more common in AR than elsewhere); May–Sep.; fruits, Jul.–Oct.

The names Maypop and Wild Apricot refer to the ripe fruits, which are delicious and can be eaten or made into a cold drink.

Yellow Passion Flower, *P. lutea* L., is a vine with similar leaves but much smaller, greenish yellow flowers; May–Sep.; fruits, Jul.–Oct.

Soapwort Gentian *Gentiana saponaria* L. [*G. puberula* Michx.]
Gentian Family Gentianaceae

Gentians are erect or sprawling leafy herbs that flower in late summer or early fall. The opposite leaves are simple. Flowers have 4 or 5 green sepals and 4 or 5 blue petals fused to form a bell-shaped corolla.

In Soapwort Gentian, several blue-violet flowers are clustered at the nodes, which also bear whorls or pairs of lanceolate leaves. Its leaves have properties like those of Soapwort (pink section), hence the name. It occurs in open moist woods, prairies, and glades of the Ozarks Plateau and Ouachitas; Sep.–Nov.

Each of these gentians has one or more features that can be used to differentiate it from the above. Pale Gentian, *G. alba* Muhl., has white or pale blue flowers; Sep., Oct. Downy Gentian, *G. puberulenta* Pringle, is a pubescent plant with dark purple, funnel-like flowers; Sep., Oct. Stiff Gentian, *G. quinquefolia* L., has wider leaves and more tubular, lilac-colored flowers; Aug.–Oct. Bottle Gentian, *G. andrewsi* Griseb., has flowers which, in full bloom, resemble buds yet to open; Aug.–Oct.

Common Morning-glory *Ipomoea purpurea* (L.) Roth
Morning-glory Family Convolvulaceae

Morning-glories belong to a principally tropical family. All in our area are vines with funnel-shaped flowers. Common Morning-glory, native to tropical Amer., is a twisting vine with pubescent stems, cordate leaves, and flowers that vary from blue/purple to pink or white. It occurs sporadically throughout the Ozarks, where it inhabits disturbed areas; Jul.–Oct.

The more common Ivy-leaved Morning-glory, *I. hederacea* Jacq., is similar but has 3-lobed leaves; Jun.–Oct. Wild Potato Vine, *I. pandurata* (L.) Mey., has cordate leaves and white flowers; Jun.–Sep.

Other morning-glories are featured in the red/orange section.

Passion Flower
Passiflora incarnata

Soapwort Gentian
Gentiana saponaria

Common Morning-glory
Ipomoea purpurea

Sand Phlox *Phlox bifida* Beck
Phlox Family Polemoniaceae

Sand Phlox forms thick, spreading mats. Lavender flowers ($\frac{2}{3}$ in. across) are 5-petaled with a prominent V-shaped notch at the end of each petal. Its small leaves are opposite and linear. It occurs on rock outcrops, slopes, and in ravines, often on sandy soils. In the Ozarks, it is apparently confined to the Ozark Plateau of MO and AR; Mar.–May.

This plant is ideal for sunny rock gardens.

Moss Phlox, *P. subulata* L., a cultivated ornamental, is also a low-growing, mat-forming phlox with bluish (or pink or white) flowers. It can be separated from Sand Phlox by its needlelike leaves and less deeply notched petals. Also called Thrift, it has become naturalized in parts of the Ozarks of AR; Apr., May.

Wild Blue Phlox *Phlox divaricata* L.
Phlox Family Polemoniaceae

This showy, branching plant is less than 2 ft. tall and has small, opposite, lanceolate leaves on slightly hairy stems. The flowers, 1 in. across, vary from light blue to pale violet. Also called Wild Sweet William, it is abundant in dry or moist woods and thickets throughout the Ozarks; Apr.–Jun.

Native Americans boiled the roots to produce a tea for the treatment of sexually transmitted diseases. The leaves were used to treat eczema and as a "blood purifer."

Other phloxes are featured in the pink section.

Jacob's-ladder *Polemonium reptans* L.
Phlox Family Polemoniaceae

Upright or partially reclining, Jacob's-ladder is a perennial that grows to about 1 ft. tall. Note the compound leaves with pointed leaflets and the small ($\frac{1}{2}$ in. across) flowers with white stamens. Also called Greek Valerian, it grows in moist woods, especially along streams. It occurs throughout the Ozarks except for OK; Apr.–Jun.

This is the only Ozarkian species of the genus. Native Americans used its pine-scented roots to make a tea for the treatment of snakebite and a wide range of digestive and respiratory ailments.

Sand Phlox
Phlox bifida

Wild Blue Phlox
Phlox divaricata

Jacob's-ladder
Polemonium reptans

Virginia Waterleaf *Hydrophyllum virginianum* L.
Waterleaf Family Hydrophyllaceae

Members of this genus have white to light blue or violet, bell-like flowers, ½ in. across, and prominent stamens that extend beyond the corolla. Both common and genus names refer to leaves that appear as if stained with water. Identification to species is based primarily on leaf shape.

Virginia Waterleaf is a smooth, 1- to 3-ft.-tall plant with large, irregular leaves with 5–7 pointed lobes. Flowers are violet to white. It occupies moist soils of woods and thickets, often on floodplains of streams, on the Ozark Plateau (MO, AR) and in the IL Ozarks; Apr.–Jul.

Browne's Waterleaf, *H. brownei* Kral & Bates, a rare plant of the w. Ouachitas, has tuberous roots and compound leaves with 9–13 leaflets; Apr., May.

Other *Hydrophyllum* species have simple maplelike leaves. Broadleaf Waterleaf, *H. canadensis* L., has flowers held well above its leaves; May–Jul. Woolen Breeches, *H. appendiculatum* Michx., has lavender or white flowers on shorter stalks and hairy leaves.

Hydrolea *Hydrolea ovata* Nutt.
Waterleaf Family Hydrophyllaceae

Hydrolea is an erect 2- 3-ft.-tall perennial. Leaves are alternate and ovate; spines are present in their axils. Bell-like purple flowers are clustered at the top of the plant. It is locally abundant in wetlands of the Ouachita Mountains, Ozark Plateau of AR, and Crowley's Ridge; less common in Ozark Plateau of MO and absent from OK and the IL Ozarks; Jun.–Sep.

Also called Hydrolea, *H. uniflora* Raf., which has a similar distribution, has lanceolate leaves and flowers arranged along the length of the stem; Jun.–Sep.

Small-flowered Phacelia *Phacelia gillioides* Brand
[*P. dubia* (L.) var. *gilioides* (Brand) GL]
Waterleaf Family Hydrophyllaceae

Related to the waterleaves (above), phacelias are similar but have narrower leaves and flowers in less dense clusters. This species is a low (12–18 in.), spreading plant with leaves (3 in. long) divided into several blunt-tipped lobes. The abundant flowers are in loose clusters. Each of the 5 petals has margins either entire (untoothed) or with short, fine hairs (as seen here). It occupies a variety of open sites in central and northcentral AR and in e. OK, but is more common in the Ozark Plateau of MO; Apr.–Jun.

Miami-mist, *P. purshii* Buckl., is a very similar plant found primarily on e. Ozark Plateau of MO. Its flowers have whitish or light violet petals with a more prominent fringe along their margins; Apr.–Jun. Hairy Phacelia, *P. hirsuta* Nutt., is found throughout most of the Ozarks. Compared to *P. gillioides,* it is a larger, stouter, and more hairy plant with unfringed petals; Apr.–Jun.

Virginia Waterleaf
Hydrophyllum virginianum

Hydrolea
Hydrolea ovata

Small-flowered Phacelia
Phacelia gillioides

Purple Phacelia *Phacelia bipinnatifida* Michx.
Waterleaf Family Hydrophyllaceae

Purple Phacelia is a 2-ft.-tall, upright biennial. Note the orange-tipped stamens that extend beyond the unfringed blue flowers, which are ½ in. across, and the segmented, coarsely toothed leaves. It is a plant of rich woods and stream banks. In our area it is found mainly on the e. Ozark Plateau of MO and AR; also in the IL Ozarks; Apr.–Jun.

Phacelia, *P. ranunculacea* (Nutt.) Const., is a similar but less common species of the same general habitats and areas. It can be distinguished from *P. bipinnatifida* by its reclining habit (weak stems), smaller size (less than 1 ft.) and smooth (vs. hairy) filaments; Apr., May.

Virginia Bluebells *Mertensia virginica* (L.) Pers.
Borage Family Boraginaceae

Plants of the borage family are known for their one-sided, rolled-up flower coils that unfurl as the flowers mature. Each flower has 5 sepals, 5 petals, and 5 stamens. The simple leaves are alternate.

Virginia Bluebells (not related to the bluebells of the Campanulaceae) is a distinctive 1- to 2-ft.-tall plant with large gray-green lanceolate leaves. Its blossoms first appear as pink buds; they open to form blue (rarely white) bell-like flowers. It often forms large colonies in alluvial soils along stream banks of the Ozark Plateau (MO, AR), and in the IL Ozarks; Mar.–Jun.

Also called Virginia Cowslip, it can be propagated for wildflower gardens by planting the 4 wrinkled seeds formed by each flower.

Wild Bergamot *Monarda fistulosa* L.
Mint Family Lamiaceae

Members of the mint family are distinctive because of their "square stems" (4-sided as seen in cross-section). Most have opposite leaves and are aromatic. Bilabiate (2-lipped) flowers, in clusters, have upper and lower lips that guard the opening into the corolla tube; inside are 2 long stamens and 2 short ones.

Flowers of this 2- to 3-ft.-tall plant vary from pink to lilac; bracts (underneath) are often lilac also. Also called Horsemint, it is a common plant of open, dry sites throughout the Ozarks; May–Aug.

This showy perennial is often cultivated. An oil extracted from its leaves has been used for the treatment of various digestive and respiratory ailments; an aromatic tea can be made from its leaves.

Purple Phacelia
Phacelia bipinnatifida

Virginia Bluebells
Mertensia virginica

Wild Bergamot
Monarda fistulosa

Hairy Skullcap *Scutellaria elliptica* Muhl.
Mint Family Lamiaceae

Skullcaps (*Scutellaria* species) are so named because of the small protuberance on the upper lip of the calyx, suggesting a cap or helmet.

Hairy Skullcap, a plant 1–3 ft. tall, has flowers in branching racemes. Both leaves and stem are rather hairy. Leaves are ovate with scalloped margins and short petioles. It occupies dry, rocky woods throughout most of the Ozarks, except for OK and Crowley's Ridge; May–Jul.

Other Ozarkian skullcaps have similar, terminal racemes of blue flowers. Downy Skullcap, *S. incana* Biehler, is a hoary plant, covered with minute hairs. Its leaves are longer, more pointed, and with dentate margins; Jun.–Sep. Heart-leaved Skullcap, *S. ovata* Hill, has cordate leaves; flowers have whitish lower lip; May–Oct.

Mad-dog Skullcap, *S. laterifolia* L., has racemes in the axils; Jun.–Oct. Smaller Skullcap, *S. parvula* Michx., is only 4–10 in. tall and has solitary flowers in its leaf axils; May–Jul.

Blue Sage *Salvia azurea* Lam.
Mint Family Lamiaceae

Blue Sage is a tall (4–5 ft.) grayish green perennial with sharply pointed, linear to lanceolate leaves attached to the stem (no basal leaves). Each sky blue flower (1 in. long) has a large flattened lower lip. It occupies limestone glades, prairies, and rocky roadsides of the Ouachita Mountains and Ozark Plateau; Jul.–Sep.

Reflexed Salvia, *S. reflexa* Hornem., is a smaller annual with blunt-tipped leaves and smaller flowers (no flattened lip). Both its flowers and its leaves angle upward; May–Oct. Lyre-leaved Sage, *S. lyrata* L., has pale blue tubular flowers in whorls along the stem. Shape of basal leaves suggest a stringed instrument, thus its name; Apr.–Jun.

White-leaved Mountain-mint *Pycnanthemum albescens* T & G
Mint Family Lamiaceae

Mountain-mints are branching plants with flat-topped clusters of small flowers that open only a few at a time. Leaves are used for identification.

This plant has whitish flowers with purple spots. Leaves are grayish green. It occurs throughout the Ozarks except in IL where it is replaced by the very similar Hoary Mountain-mint, *P. incanum* (L.) Michx. It grows in acidic soils of woods and clearings; Jul.–Sep.

The following are also widespread species. Short-toothed Mountain-mint, *P. muticum* (Michx.) Pers., has smaller clusters of flowers and larger, broader leaves; Jul.–Sep. Hairy Mountain-mint, *P. pilosum* Nutt., has somewhat narrower, hairy leaves; Jul.–Sep. Slender Mountain-mint, *P. tenuifolium* Schrad., has sharply pointed, linear leaves; Jun.–Sep.

Hairy Skullcap
Scutellaria elliptica

Blue Sage
Salvia azurea

White-leaved Mountain-mint
Pycnanthemum albescens

Downy Wood-mint *Blephilia ciliata* (L.) Benth.
Mint Family Lamiaceae

Downy Wood-mint has several tiers of blue-purple flowers separated by rows of fringed bracts. The nearly sessile leaves are downy underneath. The plant, 1–3 ft. tall, is found in dry woods, thickets, and glades throughout most of the Ozarks (less common in Ouachita Mountains); May–Aug.

The similar Hairy Wood-mint, *B. hirsuta* (Pursh) Benth., is a more hairy plant; leaves have long petioles; May–Sep.

Native Americans poulticed leaves of wood-mints for headaches.

American Dittany *Cunila origanoides* (L.) Britt.
Mint Family Lamiaceae

Dittany is a low (less than 1 ft. tall) perennial with thin, wiry stems. Its small flowers, in axils of leaves, vary from violet to white. It is found in acidic soils of dry, open woods throughout the Ozarks; Jul.–Nov.

The minty, aromatic leaves have been used to make a tea for snakebite, fevers, and headaches, and·to stimulate menstruation and perspiration. In late fall and early winter, ribbonlike, arching "frost flowers" often form around the base of the plant. This phenomenon, also know in several other Ozarkian plants, deserves further investigation.

Blue Star *Amsonia tabernaemontana* Walt.
Dogbane Family Apocynaceae

Also called Blue-dogbane, this and other *Amsonia* species are herbs with alternate leaves and sky blue, starlike flowers. Blue Star, which grows to 3 ft. tall, has flowers ½ in. across. Seed pods are held upright. Its sharply lanceolate leaves are smooth (but dull), alternate, and crowded on the stem. It grows in moist woods and along riverbanks of all regions of the Ozarks; Apr., May.

Shining Blue Star, *A. illustris* Woodson, differs by its leaves, which are shiny above, and its seedpods, which are pendant; it occurs primarily on sandbars; Apr., May. Blue Star, *A. ciliata* Walt., a limestone glade plant, is easily identified by its dull linear leaves; Apr., May. Ouachita Blue Star, *A. hubrichtii* Woodson, found on gravel bars of streams, has glossy linear leaves; Apr., May.

All 4 species are attractive plants for shady wildflower gardens. They can be grown from seed.

Downy Wood-mint
Blephilia ciliata

American Dittany
Cunila origanoides

Blue Star
Amsonia tabernaemontana

Bluets *Hedyotis caerulea* (L.) Hook [*Houstonia caerulea* L.]
Madder Family Rubiaceae

Hedyotis (formerly *Houstonia*) species, called bluets, are small plants with 4-petaled flowers, whitish or bluish. Most of the 10 Ozarkian species form a low carpet on moist ground.

Note the yellow centers of the light blue flowers (½ in. across) of this common widespread perennial. Leaves, both basal and attached to the stems, are small, sparse, and ovate. Also called Quaker Ladies, the dainty wildflowers (less than 6 in. tall) occupy moist acid soils of meadows, glades, prairies, and open woods. The plant occurs throughout the Ozarks, except for OK and Crowley's Ridge; Apr., May.

Cherokees used a leaf tea to correct bed-wetting.

Small Bluets or Star Violet, *H. crassifolia* Raf., is a common smaller (less than ¼ in. across), variable plant with smaller, usually dark blue or purple flowers that lack the yellow center. It is found in lawns and along roadsides; Feb.–Apr.

Mountain Houstonia *Hedyotis purpurea* (L.) T & G. [*Houstonia purpurea* L.]
Madder Family Rubiaceae

Its large size (to 1½ ft. tall) helps to distinguish this from the smaller *Hedyotis* species. Below the terminal cluster of pale violet tubular flowers are large paired, ovate, sessile leaves. It occupies rocky, moist woods, bluffs, and stream banks of all regions of the Ozarks; May, Jun.

Long-leaved Bluets, *H. longifloria* (Gaertn.) Hook., is a somewhat smaller (to 1 ft.) plant but with much narrower stem leaves and a basal rosette of wider leaves; Apr.–Jul. Narrow-leaved Bluets, *H. nigricans* (Lam.) Fosberg, has paired linear leaves plus clusters of smaller leaves in their axils. Flowers are both in terminal clusters and in the upper axils; May–Oct.

Sessile-leaved Monkey-flower *Mimulus ringens* L.
Snapdragon Family Scrophulariaceae

The shape of the flowers, which suggests a face, accounts for the unusual name. In this 1- to 3-ft.-tall plant, violet flowers are borne on long stalks; toothed, lanceolate leaves are sessile; stems are 4-angled. It is a wetland plant found on the Ozark Plateau of MO, OK, and n. AR, Crowley's Ridge of MO, and the IL Ozarks; Jun.–Sep.

Common Monkey-flower, *M. alatus* Ait., has similar flowers but on much shorter stalks. Leaves, which are wider, are stalked (not sessile); Jun.–Sep. The rare Monkey-flower, *M. glabratus* H.B.K., is a semiaquatic plant with small, opposite, nearly circular leaves. It is known in the Ozarks only in OK and in a few counties of the Ozark Plateau of MO; May–Oct.

Bluets
Hedyotis caerulea

Mountain Houstonia
Hedyotis purpurea

Sessile-leaved Monkey-flower
Mimulus ringens

Blue-eyed Mary *Collinsia verna* Nutt.
Snapdragon Family Scrophulariaceae

Each bicolored flower of Blue-eyed Mary has 5 lobes: 2 upper white ones and 2 lower blue ones, with the fifth concealed between. Note the toothed, sessile, lanceolate leaves in whorls. It grows along shaded stream banks of the Ozark Plateau (MO, AR, OK) and the IL Ozarks; Apr.–Jun.

Violet Collinsia, *C. violacea* Nutt., is a similar, more common and widespread plant found in somewhat drier sites. Its bicolored flowers are white (upper lobes) and violet (lower ones); Apr.–Jun.

Venus's Looking-glass *Triodanis perfoliata* (L.) Nieuwl. [*Specularia perfoliata* L.]
Bellflower Family Campanulaceae

Solitary violet-blue flowers, 5-petaled with spreading lobes, are present in the axils of nearly circular, clasping leaves. Growing to 2 ft. in height, it is a plant of prairies and waste places throughout the Ozarks; Apr.–Aug.

Tall Bellflower *Campanula americana* L.
Bellflower Family Campanulaceae

Campanula species are annual herbs with alternate, simple, lanceolate leaves with toothed margins and often bell-like (sometimes not) flowers, usually blue. Tall Bellflower is a tall (2–6 ft.) plant with both leaves and flowers along its stalk. Flowers (1 in. across) are not bell-shaped but rather composed of 5 widely spread petals; note also the long, curved style. It commonly occupies shaded, moist woods throughout the Ozarks; Jun.–Oct.

Harebell, *C. rotundifolia* L., has tiny dark blue, bell-like flowers that dangle from the ends of its branches. Leaves (except for rounded basal leaves) are linear. It is known in the Ozarks only on limestone bluffs along the Current River of Shannon Co. of southcentral MO; May–Aug.

Great Lobelia *Lobelia siphilitica* L.
Bellflower Family Campanulaceae

Related to bellflowers (above), lobelias are also herbs with simple, alternate leaves. Their tubular flowers have 2 narrower lobes above 3 wider ones. Great Lobelia, 1–3 ft. tall, is the largest of the group. The large (1 in. long) blue flowers with white stripes on the corolla tubes make identification certain. It is found in wet sites throughout; Aug.–Oct.

Other lobelias also with blue flowers: Indian Tobacco, *L. inflata* L., has smaller, swollen-based flowers in leaf axils, Jun.–Oct.; Downy Lobelia, *L. puberula* Michx., is a hairy plant with flowers in a 1-sided spike; Aug.–Oct. Pale-spiked Lobelia, *L. spicata* Lam., has small white to light blue flowers in a spike; May–Aug.

Cardinal Flower, *L. cardinalis* L., is featured in the red/orange section.

Blue-eyed Mary
Collinsia verna

Venus's Looking-glass
Triodanis perfoliata

Tall Bellflower
Campanula americana

Great Lobelia
Lobelia siphilitica

Wild Petunia *Ruellia humilis* Nutt. [*R. caroliniensis* (Walt.) Steud.]
Acanthus Family Acanthaceae

Ruellias are herbs (1–3 ft. tall) with opposite leaves and blue, trumpetlike flowers. From each slender corolla tube extend 5 flaring lobes. At the base of each flower is a pair of small leaves.

Wild Petunia is a hairy plant with long corolla tubes lacking stalks (or with very short ones). Also called Hairy Ruellia, it occurs in dry sites such as prairies, glades, and bluffs throughout most of the Ozarks; May–Oct.

Smooth Ruellia, *R. strepens* L., is a smoother plant with blue (rarely white), stalked flowers; May–Oct.

The genus *Ruellia* was named for French herbalist Jean Ruelle (1474–1537).

Water-willow *Justicia americana* (L.) Vahl
Acanthus Family Acanthaceae

This aquatic/wetland plant, 1–3 ft. tall, has slender, willowlike, opposite leaves and clusters of bicolored (lilac/white) flowers; lower lip of each is distinctively 3-lobed. It often forms large colonies in shallow water along the edges of lakes and slow-moving streams. Its range extends throughout the Ozarks except for Crowley's Ridge of AR; May–Oct.

Lance-leaved Water-willow, *J. ovata* (Walt.) Lindau, has pale purple flowers with dark violet spots on lower lip and similar but wider leaves. It occurs in swamps and wet woodlands; May, Jun.

The genus honors James Justice, 18th-century Scottish botanist.

Mistflower *Eupatorium coelestinum* L.
Aster Family Asteraceae

An overview of the aster family appears in the yellow section.

This, our only blue or blue/purple *Eupatorium*, is a 1- to 3-ft.-tall plant with flat-topped clusters of fuzzy flowers and opposite leaves. Often cultivated, it also occurs naturally in moist places, often along stream banks, probably in every county of the Ozarks; Jul.–Oct.

On rare occasions one may encounter plants with white or pinkish flowers. See other *Eupatorium* species in the white and pink sections.

The genus *Eupatorium* honors Mithridates Eupator, an early medical botanist (132–63 B.C.).

Wild Petunia
Ruellia humilis

Water-willow
Justicia americana

Mistflower
Eupatorium coelestinum

Purple Coneflower *Echinacea purpurea* (L.) Moench
Aster Family Asteraceae

Echinacea species are perennial prairie plants that occur with some frequency in the Ozarks. They are generally rough or hairy plants with lanceolate, alternate leaves and large flower heads. Rays range from yellow or lavender to purple; cone-shaped discs are spiny, thus the genus name *Echinacea*.

Purple Coneflower is a 2- to 3-ft.-tall plant with slightly reflexed, reddish purple rays. Basal leaves are petioled, ovate with prominent teeth and sharp tips; upper ones are lanceolate, nearly sessile with entire margins. It occupies prairies, thickets, and open woods throughout the Ozarks; May–Oct.

Narrow-leafed Coneflower, *E. angustifolia* DC., differs primarily by its more narrow leaves with entire margins. Rays are purple, pinkish, or white. Less common than *E. purpurea* in the Ozarks, it is apparently confined to OK; May–Oct.

Pale Purple Coneflower *Echinacea pallida* Nutt.
Aster Family Asteraceae

Pale Purple Coneflower bears flower heads with narrow, lavender, drooping rays. Leaves, mostly basal, are long and lanceolate. It occurs in glades, bald knobs, prairies, and other open, dry habitats throughout the Ozarks; May–Jul.

Plains Indians utilized roots of *Echinacea* species for a wide range of ailments. Currently over 200 pharmaceuticals containing echinacea are sold in Germany. Modern research supports the use of echinacea as an immunostimulant (stimulating the immune system, it helps to protect against infections and cancer).

Rough Blazing-star *Liatris aspera* Michx.
Aster Family Asteraceae

Blazing-stars are tall, wandlike plants with purple or pink flower heads. Leaves are alternate, linear, and often quite narrow. Identification to species usually requires close examination of the scaly bracts under the flower heads.

Rough Blazing-star grows 1–2½ ft. tall. Its purplish (rarely white) flower heads are sessile (attached to the stem directly or on very short stalks.) Note the floral bracts, which are rounded and slightly curled backward. It occupies upland glades, bald knobs, and other dry sites throughout most of the Ozarks; Aug.–Nov.

Prairie Blazing-star, *L. pycnostachya* Michx., also has sessile flowers, but its bracts are long-pointed and reflexed; Jul.–Oct. Other species have stalked flower heads. Scaly Blazing-star, *L. squarrosa* (L.) Michx., has large heads with flaring, pointed bracts; Jul.–Sep. Cylindrical Blazing-star, *L. cylindracea* Michx., has bluntly pointed bracts in a tight cylinder; Jul.–Sep.

Purple Coneflower
Echinacea purpurea

Pale Purple Coneflower
Echinacea pallida

Rough Blazing-star
Liatris aspera

New England Aster *Aster novae-angliae* L.
Aster Family Asteraceae

Named after the Greek word for star, asters have distinctive flattened flower heads. Disc flowers are yellow, changing to orange or purple; rays vary from white to blue or purple (rarely pink). Most bloom in late summer or fall. Because of frequent hybridization among the more than 20 Ozarkian species, identification to species is often difficult.

This tall (3–7 ft.) aster has numerous deep violet (or rose) rays and sticky bracts beneath. The untoothed, sessile, lanceolate leaves are attached alternately to the hairy stem. It grows in moist meadows, thickets, and stream banks sporadically in all regions of the Ozarks (but less common in OK and AR than MO and IL); Aug.–Oct.

Native Americans treated fevers and diarrhea with a root infusion of New England Aster.

These asters also have showy heads and untoothed (or nearly so) lanceolate leaves: Late Purple Aster, *A. patens* Ait., whose leaf bases almost encircle the stem, Aug.–Oct.; Smooth Aster, *A. laevis* L., whose leaf bases clasp the stem, Aug.–Oct.

Small White Aster *Aster vimineus* Lam.
Aster Family Asteraceae

This 2–5 ft. aster has purplish green stems; small, slender, untoothed leaves; and small, crowded flower heads. Ray flowers vary from whitish to light blue. It occurs in moist to wet sites such as meadows and borders of sinkholes, ponds, and swamps. Its range includes all of the regions of the Ozarks; Aug.–Oct.

White Woodland Aster, *A. lateriflorus* (L.) Britt., is a somewhat smaller plant with larger, toothed leaves. The name Calico Aster reflects the contrast between the dark purple disc flowers and light blue rays; Aug.–Oct.

Stiff Aster *Aster linariifolius* L.
Aster Family Asteraceae

One of several asters with linear leaves, Stiff Aster is a small plant (1 ft. tall). The numerous, short leaves are rigid. Each of the 1 to few flower heads is terminal. It occupies acidic soils of rocky banks and ledges and also occurs in pine-oak and oak-hickory forests of the Ouachita Mountains, Ozark Plateau, and Crowley's Ridge of MO; Aug.–Oct.

Bushy Aster, *A. dumosus* L., also with linear leaves, is a tall, branching aster bearing dozens of flower heads; Aug.–Oct.

New England Aster
Aster novae-angliae

Small White Aster
Aster vimineus

Stiff Aster
Aster linariifolius

Western Daisy *Astranthium integrifolium* (Michx.) Nutt.
Aster Family Asteraceae

One of the few daisylike spring-flowering wildflowers, Western Daisy is quite distinctive. Growing no taller than 1 ft., it has small flower heads, ½ in. across. Note the bright yellow disc, and rays that vary in color from lavender on the outside edge to white at the center. It is seen on rocky soils of roadsides, glades, and prairies; also on alluvial soils of stream banks. Its range includes the Ouachitas, most of the Ozark Plateau of AR and OK, but only a few counties of sw. MO: McDonald, Barry, Stone, and Taney; Mar.–Jun.

Tall Ironweed *Vernonia gigantea* (Walt.) Trel. ex Bran & Cov. [*V. altissima* Nutt.]
Aster Family Asteraceae

Ironweeds are tall (to 10 ft.) plants that somewhat resemble Joe-Pye-weeds (pink section) but have more open clusters of flower heads with deeper colors (violet-purple vs. pinkish purple). Leaves are more slender, sessile, and alternate. The name refers to their tough stems. As hybridization among the half-dozen Ozarkian species is common, identification to species is often difficult.

Tall Ironweed has rose-purplish (rarely white) flowers and narrowly lanceolate leaves that are downy underneath. Bracts (modified leaves surrounding the flower heads) are rounded. It occupies moist soils near streams, swamps, and low woods and thickets throughout the Ozarks; Aug.–Oct.

Baldwin's Ironweed, *V. baldwinii* Torr., is distinguished from Tall Ironweed by its broad, sharp-tipped, reflexed bracts. Also, it grows mainly in drier sites and flowers earlier; May–Sep. Great Ironweed, *V. arkansana* DC., has slender, pointed bracts; Jul.–Sep.

The genus commemorates the English botanist William Vernon, who traveled in N. Amer. during the late 17th century.

Missouri Ironweed *Vernonia missurica* Raf.
Aster Family Asteraceae

Smaller (to 4 ft.) than other ironweeds, *V. missurica* also has smaller, barely toothed, less elongated, and more hairy leaves. The less numerous flower heads have overlapping bracts, each with a green or purple midvein, which is lacking in those of other species. It occurs in moist sites throughout the Ozarks; Jul.–Sep.

Letterman's Ironweed, *V. lettermanii* Engelm., is a plant of similar size but with numerous needlelike leaves (3 in. long). It occupies rocky sites in the Ouachita Mountains; Aug.–Sep.

Native Americans used a root tea of various *Vernonia* species for childbirth pain and to regulate menses.

Western Daisy
Astranthium integrifolium

Tall Ironweed
Vernonia gigantea

Missouri Ironweed
Vernonia missurica

Pokeweed *Phytolacca americana* L.
Pokeweed Family Phytolaccaceae

Poke or Pokeweed is an interesting plant. Growing to a height of nearly 10 ft., it has red stems and long, smooth, oblong to lanceolate leaves. Long racemes of numerous tiny greenish white flowers produce dark purple berries; flowers and berries are often on a plant at the same time. It is common in disturbed sites throughout the Ozarks; May–Oct; fruits, Jun.–Nov.

"Poke salad" is a traditional Ozark food. Leaves, collected in the spring, must be cooked in several waters before eaten as a potherb. Stems, if eaten, can cause serious poisoning. Seeds are also considered poisonous, but the purple berry juice has been used as a dye and food coloring as well as a substitute for ink. Native Americans used a berry tea for arthritis, rheumatism, and dysentery.

Black Gum *Nyssa sylvatica* Marsh.
Tupelo Family Nyssaceae

Black Gum or Sour Gum (not related to Sweet Gum) is a tall tree (to 90 ft.) of moist soils. Small greenish flowers are borne in separate clusters: staminate (male) and pistillate (female). From the latter are formed the dark blue fruits (½ in. long) seen here. Its shiny leaves, 2–5 in. long, turn to a brilliant crimson, making the tree a spectacular autumn sight. Black Gum occurs throughout the Ozarks in a variety of dry to wet habitats; Apr.–Jun.; fruits, Sep.–Oct.

The berries are eaten by a variety of birds and other wildlife such as deer. The wood is used for pulpwood in papermaking and also a variety of items where a very hard, tough wood is required.

Swamp Tupelo, *N. aquatica* L., has larger leaves (5–7 in. long) with toothed margins, and larger fruits. Growing in wetlands, it has swollen trunk bases like those of Baldcypress; Mar.–Apr.; fruits, Sep.–Dec.

Beautyberry *Callicarpa americana* L.
Verbena Family Verbenaceae

Also called French Mulberry, this spreading shrub (to 5 ft.) has arching branches. Clusters of tiny lavender flowers surround the petioles of the opposite leaves at regular intervals. Margins of the sticky opposite leaves vary from serrate to dentate (seen here). The clusters of violet-purple berries appear almost too gaudy to be natural. Not found in MO or sw. IL, it is widespread and relatively common in the Ouachita Mountains (AR, OK), Ozark Plateau of AR, and Crowley's Ridge of AR. It occupies a variety of sandy and rocky sites; Jun.–Aug; fruits, Aug.–Nov.

The berries are eaten by a variety of songbirds and other wildlife. The shrub is planted as an ornamental. Native Americans used teas made from roots, leaves, and berries for a variety of ailments.

Pokeweed
Phytolacca americana

Black Gum
Nyssa sylvatica

Beautyberry
Callicarpa americana

Green / Brown

Jack-in-the-pulpit *Arisaema triphyllum* (L.) Schott [*A. atrorubens* (Ait.) Blume]
Arum Family Araceae

Arums are characterized by a thick flowering stalk, the spadix, surrounded by a cylindrical spathe, a modified leaf.

In this striking wildflower, 12–18 in. tall, the spadix, or "jack," is enclosed by the striped spathe with a flap above that forms the "pulpit." Each plant has 1 or 2 large compound leaves, each with 3 leaflets. It is found in rich, moist woods throughout our area; Apr.–Jun; fruits, Aug.–Oct.

During any given season, each plant is either male, producing only staminate flowers, or female, with pistillate flowers.

Green Dragon *Arisaema dracontium* (L.) Schott
Arum Family Araceae

Compared with Jack-in-the-pulpit (above), this plant is taller (to 4 ft.), has a longer, thinner spathe with a spadix that extends far beyond the spadix, and leaves divided into 5–15 leaflets. It is also common throughout the Ozarks but in generally more moist situations; Apr.–Jun.

The only other Ozark representatives of the arum family are Arrow Arum, *Peltandra virginica* (L.) Schott & Endl., which has huge arrow-shaped leaves, Apr.– Jun.; and Sweet Flag, *Acorus americanus* Raf., with greenish fingerlike spadixes (no spathes) attached at an angle to swordlike leaves, May–Jul. Both species occur in wetlands.

Common Cat-tail *Typha latifolia* L.
Cat-tail Family Typhaceae

Cat-tails are tall plants with compact cylinders of tiny flowers. They grow in shallow water. The staminate (male) flowers are located above the pistillate (female) ones. This species is the only common cat-tail of the Ozarks. In the plants seen here, the staminate flowers, directly above the pistillate flowers still present, have already been shed. It is found in all major regions of the Ozarks; May–Jul.

Several parts of cat-tails are edible: the young spiny shoots can be used like asparagus; the pollen, as flour; and the rootstocks, as a starch source in winter.

False Aloe *Agave virginica* L. [*Manfreda virginica* Salisb.]
Agave Family Agavaceae

Thick leaves form rosettes at the base of the tall (3–6 ft.) flowering stalks of False Aloe. The inconspicuous greenish yellow flowers are believed to be pollinated by moths attracted to the fragrance released at night. Also called Rattlesnake-master, it is found in limestone glades and dry, sandy woods throughout our area; Jun.–Aug.

Jack-in-the-pulpit
Arisaema triphyllum

Green Dragon
Arisaema dracontium

Common Cat-tail
Typha latifolia

False Aloe
Agave virginica

Wild Yam *Dioscorea quaternata* J. F. Gmelin [*D. villosa* L.]
Yam Family Dioscoreaceae

Dioscorea species are rather common, nonwoody, climbing or sprawling vines with small greenish flowers; they lack tendrils.

Wild Yam has whorled cordate leaves, hairy beneath and with prominent veins. Its inconspicuous flowers are followed by clusters of brown, 3-winged, inflated fruits that are retained for much of the winter. It occupies thickets, edges of woods, and other, often dry, places throughout; Apr.–Jun.; fruits, Sep.–spring.

Green Wood Orchid *Platanthera clavellata* (Michx.) Luer [*Habenaria clavellata* (Michx.) Spreng.]
Orchid Family Orchidaceae

Platanthera species are orchids with small flowers, usually with fringed petals, arranged into more or less showy spikes. This 6- to 18-in.-tall plant bears greenish or yellowish flowers, each with a short, blunt, unfringed petal and a curved spur. An uncommon plant of swamps and springs, it is known from Fulton Co. of n. AR and several counties of Crowley's Ridge of AR and MO; Jul., Aug.

Pale Green Orchid, *P. flava* (L.) Lindley, is a similar but taller (to 24 in.) plant. The presence of several erect, lanceolate leaves surrounding the base of the flowering stalk distinguishes it from *P. clavellata;* May–Sep.

Putty-Root *Aplectrum hyemale* (Muhl. ex Willd.) Torr.
Orchid Family Orchidaceae

The single, leathery basal leaf at the base of each flowering stalk is the most useful means of identifying this wildflower: it is dark blue-green above and deeply ridged and purplish beneath. As the leaf withers, the flowering stalk (to 1 ft. tall) develops. The small flowers (1 in. long) are usually purplish green. It is a plant of rich, moist woods, often in ravines or along streams. It occurs along the e. edge of the Ozark Plateau in MO. In AR it is seen in the s. Ozark Plateau and the n. Ouachita Mountains; also in e. OK and IL Ozarks; May, Jun.

Crane-fly Orchid *Tipularia discolor* (Pursh) Nutt.
Orchid Family Orchidaceae

Like Putty-root (above), this inconspicuous plant also has a single oval leaf, purplish beneath, that persists throughout the winter. Each small purplish brown flower, ½ in. wide, has a slender spur that extends back toward the 1- to 2-ft.-tall spike. It is found principally in moist deciduous forests. Known only from AR in our area, it has been reported more often from the Ouachita Mountains but is seen also in the s. Ozark Plateau; Jul.–Sep.

Wild Yam
Dioscorea quaternata

Green Wood Orchid
Platanthera clavellata

Putty-root
Aplectrum hyemale

Crane-fly Orchid
Tipularia discolor

Green Violet *Hybanthus concolor* (T. F. Forst.) Spreng.
Violet Family Violaceae

Although a member of the violet family, this coarse plant bears little apparent resemblance to violets (*Viola* species). The small (¼ in. long) greenish flowers are arranged singly in the axils of the lanceolate leaves. Each flower does have a clublike pistil, typical of the family. Green Violet grows to a height of 1–3 ft. in moist limestone soils throughout the Ozarks but is less common in the Ouachitas and Crowley's Ridge of AR than elsewhere; Apr.–Jun.

This is the only Ozarkian *Hybanthus* species.

Blue Cohosh *Caulophyllum thalictroides* (L.) Michx.
Barberry Family Berberidaceae

This distinctive smooth perennial (1 to 3 ft. tall) has bluish green leaves divided into 3 leaflets. Its small greenish yellow to brown flowers are arranged in a terminal raceme; berries that follow are dark blue. It occurs in rich woods of ravines and valleys of the Ozark Plateau (more common in MO than AR and OK) and the IL Ozarks; Mar.–May.

Native Americans pulverized the roots and used them to treat rheumatism, bronchitis, and menstrual cramps; they were also used as an aid in childbirth. Modern herbalists use the roots to treat these same ailments. There is scientific evidence that the plant is effective against rheumatism and that it has antispasmodic properties, but the berries are poisonous.

Buffalo Clover *Trifolium reflexum* L.
Pea Family Fabaceae

Like other clovers, this plant has leaves divided into 3 rounded leaflets and small pealike flowers in tight heads. A low plant (6 in. tall), it has tiny flowers that appear to be white/green, but closer examination reveals them to be white with purple or reddish petals, turning brownish as they mature. It occupies acidic soils of glades, fields, and open woods throughout the Ozarks; May–Aug.

Other clovers, escapes from cultivation, are identified by their flower color. Red Clover, *T. pratense* L., has pink flowers; Crimson Clover, *T. incarnum* L., red; White Clover, *T. repens* L., white. All are important as forage crops; May–Oct.

The rare Running Buffalo Clover, *T. stoloniferum* Eat., resembles common White Clover but is a somewhat larger plant with prominent stipules. More common in the 19th century than now, it is seen along old buffalo trails; May–Aug.

Green Violet
Hybanthus concolor

Blue Cohosh
Caulophyllum thalictroides

Buffalo Clover
Trifolium reflexum

American Columbo *Frasera caroliniensis* Walt.
[*Swertia caroliniensis* (Walt.) Ktze.]
Gentian Family Gentianaceae

A long-lived perennial, American Columbo produces a basal rosette of large (to 1 ft. long) oblong leaves for several years before "bolting" (forming a flower stalk), after which the entire plant dies. The stalk may reach a height of 8 ft. or more and bear dozens of flowers such as those seen here. Note the purple spots on the greenish petals; flowers are about 1 in. across. This striking plant is found on rocky slopes or rich woods, especially those of the Ouachitas, less often elsewhere in the Ozarks; May–Jun.

False Nettle *Boehmeria cylindrica* (L.) Sw.
Nettle Family Urticaceae

The nettle family includes coarse weeds of waste places. Their simple leaves have toothed margins, and their small, greenish flowers occur in spreading or upright clusters. Many have stinging hairs.

False Nettle is so called because it lacks stinging hairs. The 1- to 2-ft.-tall perennial has opposite, lanceolate leaves with dentate margins. Unisexual flowers of both types (staminate and pistillate) occur together in erect axillary spikes. It occurs in wet meadows and along stream banks throughout the Ozarks; Jun.–Oct.

Other Ozarkian nettles are covered with hairs that cause an intense burning sensation for up to an hour when they come in contact with the skin. Stinging or Wood Nettle, *Laportea canadensis* (L.) Wedd., is a similar and also widespread plant but has horizontal, lacy, branching flower clusters. Leaves are alternate and broader. Found in moist woods throughout the Ozarks, it often occurs in large stands; May–Aug. Nettle, *Urtica chamaedryoides* Pursh, resembles *L. canadensis* but has opposite leaves and smaller flower clusters in the leaf axils. It occurs throughout AR and in extreme s. MO; Apr.–Sep.

Green-flowered Milkweed *Asclepias viridis* Walt.
[*Asclepiodora viridis* (Walt.) Gray]
Milkweed Family Asclepiadaceae

This 2- to 3-ft.-tall plant has thick alternate leaves with short petioles. Also called Antelope-horn Milkweed, its flowers are somewhat different from others of the family: they are larger and their petals (green) spreading vs. reflexed; crown (upper portion) is purplish green. It occurs in rocky, usually basic soils, throughout the Ozarks; May, Jun.

Green Milkweed, *A. viridiflora* Raf., has variable opposite leaves and green flowers more like those of other milkweeds; May–Aug. Prairie Milkweed, *A. longifolia* Michx., has long, narrowly lanceolate leaves, and pinkish green flowers arranged in conspicuous, almost spherical umbels; May–Aug.

American Columbo
Frasera caroliniensis

False Nettle
Boehmeria cylindrica

Green-flowered Milkweed
Asclepias viridis

Spice-bush *Lindera benzoin* (L.) Blume
Laurel Family Lauraceae

The small flowers, in dense clusters, appear on the twigs of this aromatic, 6- to 12-ft.-tall deciduous shrub. Leaves that follow are somewhat oval with sharp tips. Fruits are shiny, bright red, elongated berries. Spice-bush occurs in moist soils of woods and thickets throughout the Ozarks; Mar.–May; fruits, Aug.–winter.

This shrub in fruit is featured in the red/orange section.

Sassafras, *Sassafras albidum* (Nutt.) Nees, of the same family, is a tree with similar flowers but in larger, rounded clusters. Its leaves, which appear after the flowers, are of several different shapes on the same tree; Apr. Sassafras tea, made from the root bark, is a traditional Ozarkian spring tonic and "blood purifier." However, it has been banned by the FDA because safrole, its essential oil, has some potential as a carcinogen.

Wafer-ash *Ptelea trifoliata* L.
Rue Family Rutaceae

A misnamed plant (not related to ashes, *Fraxinus*), this aromatic shrub or small tree (10–20 ft. tall) has leaves with 3 leaflets. Small green flowers are in cymes; the waferlike samaras (fruits) that follow are flat, circular, and 2-seeded. Wafer-ash is found on limestone soils of glades, woods, and thickets of the Ouachitas and Ozark Plateau, less commonly on Crowley's Ridge and the IL Ozarks; Apr.–Jun.

Other names for the plant are of note. Skunk-bush alludes to its disagreeable odor when flowers or bark are crushed. Hop-tree refers to the use of its aromatic, bitter roots as a substitute for hops and for quinine.

American Smoke-tree *Cotinus obovatus* Raf. [*C. americanus* Nutt.]
Cashew Family Anacardiaceae

Related to Poison Ivy and Sumac (see red/orange section), American Smoke-tree is a small, irregularly shaped tree. In spring its large lacy clusters of small reddish brown flowers create the effect of smoke when viewed from a distance. Leaves, oval to obovate and usually wider toward the end, turn a brilliant reddish orange in autumn, giving the plant a distinctive appearance at that season also. Believed to have once had a wider range, it is now a relatively uncommon tree of limestone bluffs, glades, and bald knobs of the northcentral Ozark Plateau of AR and (less often) that of southcentral MO (rare in e. OK); May.

The wood is used for fence posts and formerly as a source of an orange-colored dye. The small tree is planted as an ornamental.

Spice-bush
Lindera benzoin

Wafer-ash
Ptelea trifoliata

American Smoke-tree
Cotinus obovatus

Winter Grape *Vitis vulpina* L. [*V. cordifolia* Michx.]
Grape Family Vitaceae

The Ozark flora includes 8 *Vitis* species, woody vines that climb by tendrils high into trees. Leaves are large, more or less lobed, and often toothed. Bark is usually shreddy. Panicles of greenish flowers are followed by autumn fruits, which are eaten by wildlife and humans.

This species has unlobed leaves each with a prominent V-shaped sinus at its base; teeth along margin are wide, not sharp; both sides of the blade are green and glossy. Bluish black fruits are sour at first, becoming sweet after frost. Among the most common of our grapes, it is found along streams and other wet places throughout; May, Jun.; fruits, Sep.–Oct.

River-bank Grape, *V. riparia* Michx., is a less common but also variable species. Its leaves, as compared to those of Winter Grape, are duller and have larger, coarser, sharper teeth and a U-shaped sinus. Distinct from other grapes, Muscadine, *V. rotundifolia* Michx., has very large rounded (sycamore-like) leaves, bark that doesn't readily shed, and larger fruits (sweet and more tasty) with a thick skin; May, Jun.; fruits, Sep.–Oct.

Wahoo *Euonymus atropurpureus* Jacq.
Staff-tree Family Celastraceae

Wahoo is a large shrub with bright green twigs; leaves are opposite and elliptical with long, pointed tips and serrate margins. Flowers, ¼ in. across, are borne on long stalks in clusters of more than a dozen; each has 4 purplish brown petals and, in the center, a prominent disc that conceals the ovary. Fruits are smooth, rose-colored, and 4-lobed. The name Eastern Burning Bush alludes to its autumn appearance, which includes reddish leaves and fruits. It occurs in open woods, along streams, and in thickets throughout the Ozarks; Apr.–Jun; fruits, Sep.–Oct.

Running Strawberry Bush, *E. obovatus* Nutt., is a vine; its flowers have 5 greenish white petals. In the Ozarks it is known only from Stone and Madison Cos. of AR; Apr.–Jun; fruits, Sep. Strawberry Bush, *E. americanus* L., is featured in fruit in the red/orange section.

Carolina Buckthorn *Rhamnus caroliniana* Walt.
Buckthorn Family Rhamnaceae

Note the prominent venation of the thick, leathery, ovate leaves of this thornless shrub or small tree (to 25 ft.). The small greenish white flowers (¼ in. across), with parts in 5s, produce berries that change from bright red to purplish black as they mature. It occurs primarily in limestone areas throughout the Ozarks; May, Jun.; fruits, Sep.–Nov.

Lance-leaf Buckthorn, *R. lanceolata* Pursh, is a smaller shrub (to 6 ft.). Its leaves have serrate margins and more sharply pointed tips. Flower parts are in 4s; Apr.–Jun; fruits, Jun.–Aug.

Winter Grape
Vitis vulpina

Wahoo
Euonymus atropurpureus

Carolina Buckthorn
Rhamnus caroliniana

APPENDIX 1

Glossary

Abiotic factors. The nonliving factors of an ecosystem, such as temperature, light, and water.

Angiosperms. The largest (250,000 described species) group of plants; they reproduce by flowers, fruits, and seeds.

Annual. A plant that lives only a single growing season.

Azonal soils. Soils lacking horizontal layers; those that were carried to their present site by wind, water, or gravity.

Biodiversity. The total number and variety of species of a particular ecosystem or biotic community.

Biotic factors. Influences exerted by other organisms on a particular organism of an ecosystem.

Bog. An acidic wetland ecosystem resulting from the decomposition of peat or other organic materials.

Calcareous. Refers to basic soils formed from calcium-containing rocks; example: limestone.

Canopy. The uppermost layer of trees of a forest.

Climax ecosystem. In traditional ecological theory, a stable ecosystem in equilibrium with its environment, especially climate.

Consumers. Organisms (primarily animals) of an ecosystem that obtain their food either directly or indirectly from producers (plants and algae).

Deciduous forest. A forest in which most of the trees lose their leaves each fall.

Decomposers. Microorganisms (bacteria and fungi) that break down the bodies of organisms, making the resulting simpler substances available for recycling.

Dicots. One of the two major groups of flowering plants; characterized by flower parts in 4s or 5s and net-veined leaves.

Dolomite. A brittle sedimentary rock composed of calcium magnesium carbonate.

Dominant species. The one or more species of an ecosystem or community that, because of their size or numbers, exert a major influence.

Ecological succession. The process of regeneration by which, after a disturbance to an ecosystem, a permanent or climax ecosystem eventually results.

Ecosystem. A local unit of nature including both living and nonliving (soil, water, light, temperature) factors; e.g., pond, coral reef, bog, field, forest.

Exotic. A plant or animal occurring outside its native area.

Flora. The collective plant species of an area.

Floristic. Referring to the total listing of plants of an area without regard to their abundance.

Generic name. The first part of a scientific name; e.g., *Iris* is the generic name for *Iris cristata*.

Herb. A plant with nonwoody (as compared to shrubs and trees with woody) tissues; also applied to any plant that has a particular practical use, such as for medicine, dyeing, or flavoring.

Herbarium. A collection of dried, pressed plants maintained by a university or other research institution.

Horizons (soil). Horizontal strata of soils as seen in a soil profile (vertical cut).

Hydric. Refers to a wet habitat or environment.

Inflorescence. A cluster of individual flowers on a single stalk (peduncle).

Loess. A fine, light-colored soil carried by wind and deposited a distance from its origin.

Mesic. An environmental condition intermediate between wet (hydric) and dry (xeric); i.e., moist.

Mesophyte. A plant adapted to mesic (moist) conditions.

Microclimate. The climate of a particular local site such as a cave or the south side of a hill; each ecosystem possesses numerous microclimates.

Microhabitat. A distinctly different subdivision of a more general habitat.

Monocots. One of the two major groups of flowering plants; characterized by flower parts in 3s and parallel-veined leaves.

Old-growth forest. A forest, although not necessarily a virgin one, that approximates a climax forest; one that has not been disturbed for centuries.

Organism. Any individual living thing, be it a plant, animal, or microorganism.

Peat. The highly organic material resulting from the decomposition of plants, especially sphagnum moss.

Perennial. A plant that lives an indefinite number of years.

Producers. Organisms of an ecosystem, primarily plants and algae, that produce (by photosynthesis) the food for that ecosystem.

Reflexed. Refers to plant parts, especially sepals or petals, that are bent downwards towards the stem.

Rhizome. Underground, usually horizontal, stem of a plant.

Saponins. Chemicals, produced by some plants, that form a lather when shaken with water.

Secondary forest. A forest that has regrown after logging or other disturbances.

Species. A particular kind of plant, animal, or microorganism; members of a given species typically breed within the group but not outside it.

Successional ecosystems. Ecosystems that are in the process of undergoing ecological succession.

Tepals. Sepals and petals collectively; outer two whorls of a flower.

Terete. Refers to a leaf or other plant part that is round in cross-section.

Vegetation. The collective plant life of a given area.

Weed. An organism, usually a plant, that thrives in disturbed sites; a plant growing in a place where it is not desired.

Xeric. Refers to a dry habitat or environment.

APPENDIX 2

Ozark Natural Areas

Fortunately, the Ozark region abounds with protected areas that are available for natural history studies. Described here are some representative sites where a diversity of Ozark plants and associated biota can be experienced. Many of the wildflower photographs in this book were taken in the areas described.

The natural areas described are arranged in a generally counterclockwise fashion (except those of Oklahoma), beginning in southwestern Illinois (see map).

Illinois

These natural areas of southwestern Illinois are near Murphysboro and Carbondale (home of Southern Illinois University), where there are ample lodging and camping facilities. Maps and other information about these and other sites of interest are available at the Shawnee National Forest Station in Murphysboro. Nearby Jonesboro can also be used as a home base.

LARUE–PINE HILLS ECOLOGICAL AREA

From Murphysboro take Hwy. 149 west to Hwy. 3; go south past the village of Wolf Lake. Turn left on FR 236, which takes you to the top of Pine Hills. Continuing northward, you will pass through Allen's Flat and McGee Hill before reaching LaRue–Pine Hills Ecological Area. This unit of the Shawnee National Forest is noted for its unusually high diversity of plants and animals within a relatively small (3,000-acre) area. It has high ridges, steep bluffs, swamps, and bottomland forests. More than 1,150 plant species have been reported; 13 are state endangered species.

LaRue Rd. (FR 345) separates a 350-ft.-high bluff from LaRue Swamp. Because many reptiles and amphibians have been killed as they migrated across the road, it is now closed to vehicular traffic for several weeks in spring (usually in April) and fall (October). Cottonmouths and other poisonous snakes are in abundance; one should be wary and take reasonable precautions.

TRAIL OF TEARS STATE PARK

This preserve is in the southern section of the Illinois Ozarks. Leading eastward from the village of Ware near the LaRue area is Hwy. 146. It passes the state nursery on the way to Trail of Tears State Park, a distance of about 5 miles. Within the park is Ozark Hills Nature Preserve, a 222-acre area where one can see Cucumber Magnolia, Red Buckeye, Mountain Azalea, and many herbaceous species typical of Ozarkian mixed mesophytic forests. An unpaved loop road accesses the area.

FULTS HILL PRAIRIE NATURAL PRESERVE

This 532-acre natural area is on Hwy. 3 (Bluff Rd.), about 5 miles north of Hwy. 155. A small sign and parking lot are on the right. Lying within the northern section of the Illinois Ozarks, it includes an excellent example of a rare unglaciated loess hill prairie. From the parking lot, take the trail northward, which leads you upward through a mixed mesophytic forest that opens into a south-facing limestone glade. Continuing, you will soon reach the hill prairie with grasses including Big and Little Bluestem, Indian Grass, and Side-oats Grama. Also present are several species each of goldenrods, blazing stars, asters, and coneflowers.

From here one may return south on Hwy. 3 to the town of Modoc, where a ferry (small charge) takes you across the Mississippi River to historic St. Genevieve, Missouri. The St. Francois Mountains area, described below, is only an hour's drive west of St. Genevieve.

Missouri

Roughly half of the Ozark Plateau lies in central and southern Missouri, where the following natural areas are located.

TAUM SAUK MOUNTAIN STATE PARK

Located in the heart of the St. Francois Mountains of the northeast Ozarks, this natural area is of both geological and botanical interest. Taum Sauk Mountain, at an elevation of 1,772 ft., is the highest point in the state. From Ironton take Hwy. 21 south; after 2 miles, turn right on MO CC, which leads to the park. In addition to a short trail that leads to the high point, there is a longer one (3 miles) to Mina Sauk Falls, the highest waterfall (132 ft.) in the state. Both are recommended for their pine and oak forests, glades, and savannas typical of the region.

Also on MO CC is Russell Mountain Trail, a short section of the Ozark Trail planned to extend from St. Louis to western Arkansas, a distance of over 500 miles. Also nearby, and also a part of the St. Francois Mountains Natural Area, are Elephant Rocks State Park and Johnson's Shut-ins State Park. Lodging and dining facilities are available in Ironton; camping, in the state parks.

BIG SPRING AND BIG SPRING PINES NATURAL AREAS

From Van Buren in southeastern Missouri, Hwy. 103 leads south 4 miles to Big Spring, one of the largest Ozarks springs (daily discharge: 276 million gallons). Within the 17-acre natural area are numerous species of aquatic and wetland plants.

Just south of this area is Big Spring Pines Natural Area, a larger (345 acres) park that includes outstanding pine-oak forests with their associated wildflowers. Among the wetland plants is the uncommon Heart-leaf Plantain (*Plantago cordata*), found along several small springs. This above area is within the Ozark National Scenic Riverways, which extends along the Current and Jacks Fork Rivers.

Lodging and dining are available at Van Buren, where is also located an office of the Scenic Riverways. Books, maps, and information are available there. Camping is available at Big Spring Park. A section of the Mark Twain National Forest is nearby.

SPRINGFIELD CONSERVATION NATURE CENTER AND WILSON'S CREEK NATIONAL BATTLEFIELD

In Springfield is this 80-acre natural area, which includes woods, glades, a lake, and hiking trails. Also there is a museum/interpretive building. Numerous wildflowers can be seen, including the rare Trelease's Larkspur. From the city, go west on Glenstone Ave. (MO Business Rt. 65) and follow the signs. Camping facilities and a wide range of lodging and dining facilities are available in Springfield.

Southwest of Springfield is Wilson's Creek National Battlefield. Although Wilson's Creek is primarily a historical park, many acres of woods and prairies are protected within its boundaries. From Springfield take Hwy. 60 west to Hwy. 22, which takes you south to the visitors' center.

RUTH AND PAUL HENNING CONSERVATION AREA

Just an hour's drive south of Springfield, via Hwy. 65, is the famous entertainment city of Branson. Both the Henning Area and the adjacent White River Balds Natural Area can be reached by taking Hwy. 76 west from downtown Branson a few miles, until you see a sign pointing north to the Henning Conservation Area. The most notable plant communities of the Henning 1,534-acre area are the extensive glades, locally called bald knobs, made famous by Harold Bell Wright's book *Shepherd of the Hills*. Among the typical herbaceous glade plants are Yellow Coneflower, Missouri Evening-primrose, and Blue False Indigo. Uncommon trees include Ashe Juniper and Smoketree. During warm months naturalist-led wildflower walks leave from the visitors' center.

Arkansas

In this state are located major portions of both the Ozark Plateau and the eastern Ouachita Mountains.

BAKER PRAIRIE

From Branson, Missouri, take Hwy. 65 south to Harrison, Arkansas, a drive of only about one hour. From Harrison take Hwy. 43 west for 2 miles, then Hwy. 397 1 mile north on Hwy. 392, right on Goblin Drive to Harrison High School. The prairie is just beyond the school on both sides of the road.

Designated in 1992 as a state natural area and owned by the Arkansas Natural Heritage Commission, 71-acre Baker Prairie is a remnant of the once vast tallgrass prairie of the midcontinent. Because of the diversity of its plants, it is especially recommended. Although the largest number of species are usually in flower in late summer, any time spring through fall is a good time to visit.

BLANCHARD SPRINGS CAVERNS AND OZARK NATIONAL FOREST

From Harrison, Mountain View can be reached via Hwys. 65 and 66, a drive southeast of about 2 hours (not counting botanizing along the highway!). It is billed as the folk musical capital of the world and is a popular resort; nearby is the Ozark Folk Center.

Blanchard Springs is reached by traveling a few miles west on Hwy. 14 from Mountain View. The visitors' center is a good place to orient yourself to Blanchard Springs and the local section of the Ozark National Forest. An extensive selection of maps and books on the Ozarks is available there. Within the area are numerous trails that access streams and springs as well as pine and pine-oak forests. There is a wide diversity of wildflowers typical of the Boston Mountains, this portion of the Arkansas Ozark Plateau.

There are several local campgrounds in the Ozark National Forest.

PETIT JEAN STATE PARK AND MT. MAGAZINE STATE PARK

Located in central Arkansas, along the northern edge of the Ouachita Mountains, Petit Jean State Park is among the finest in the state. From Conway it can be reached by traveling west via Hwys. 60, 113, and 154.

Among the many hiking trails is Cedar Creek Falls Trail. Starting at Mather Lodge, the 2-mile trail takes you into Cedar Creek Canyon. Along the more moist, north-facing slopes are Red Maple, Sweet Gum, and Red and White Oak; along the drier, south-facing slopes are Shortleaf Pine, Post and Blackjack Oak and various species of hickory. In addition to the more common wildflowers associated with these forests, there are also Large Yellow Lady's-slippers.

The focal point of Petit Jean State Park is rustic Mather Lodge. Available in the lodge are rooms and a dining room. Cabins are nearby. There is also a campground and an airport in the park.

An hour's drive west, via Hwys. 154, 10, and 309, is Mt. Magazine State Park. Recently developed, it contains the highest point in the Ozarks (2,753 ft.). Its plant communities and flora are similar to those of Petit Jean State Park.

VILLAGE CREEK STATE PARK AND CROWLEY'S RIDGE STATE PARK

Within Village Creek State Park of eastern Arkansas, you can observe both the unique geology (including thick loess deposits) and flora characteristic of Crowley's Ridge. From I-40 near Forrest City, take Hwy. 284 north 13 miles to the park.

Five trails lead through the mixed mesophytic forests of southern Crowley's Ridge. In addition to Tulip-tree, there are several other plants not otherwise found west of the Mississippi River. Other plants are at their easternmost extent here. My favorite trail is the loop trail that begins near the visitors' center parking lot. Wetland wildflowers surround Lake Austell and Lake Dunn. Check with a naturalist for rare or unusual plants in flower.

A campground is located in the park. Lodging is available at Forrest City and also at the town of Wynne 10 miles north of the park.

Crowley's Ridge State Park has similar forests and facilities and also cabins for overnight stays. It is reached via Hwy. 141 north of Jonesboro.

HOT SPRINGS NATIONAL PARK

Lying within the southern Ouachita Mountains of westcentral Arkansas, this unique 1,016-acre park includes forested hills that surround downtown Bathhouse Row. From Little Rock take I-30 south, then Hwy. 70 west to Hot Springs.

Among the hills of the park are Sugar Loaf, Indian, North, and West Mountains; the latter has an elevation of 1,320 ft. The forests of the park were recently studied by Dale and Ware (see bibliography), who compared them to other oak-hickory forests. They found shortleaf pine to be codominant with oak and hickory, especially on south-facing slopes.

Camping is available in the park as well as varied lodging facilities in the city.

Oklahoma

Even though central and western Oklahoma was originally covered by prairies, oak-hickory forests of the Ozark Plateau extend westward from Arkansas into northeastern Oklahoma. Likewise, those of the Ouachita Mountains extend into the southeastern part of the state.

TALIMENA SCENIC BYWAY AND ROBERT S. KERR MEMORIAL ARBORETUM

Queen Wilhelmina State Park of southwest Arkansas, on Hwy. 88, just west of Mena, is a convenient starting point for the exploration of the Oklahoma Ouachita Mountains. The park, with an excellent lodge ("Arkansas's Castle in the Sky"), also has a campground. Atop the second highest mountain of the state, it is an ideal place for enjoying fall foliage (but make reservations months in advance).

From Queen Wilhelmina State Park, take Hwy. 88 (Talimena Byway) westward. After a few miles you will reach the State Line Historical Site. The path from the parking lot to the survey marker offers a chance to sample typical Ouachita flora. Continuing westward on the Byway for several miles, you will note stunted post oak and blackjack oak trees composing the forest on both sides of the road. The small stature of the trees, known as an "elfin forest," is explained by the shallow soil combined with their exposure to extreme cold and hot temperatures.

A short distance westward is the entrance to the Nature Center of Robert S. Kerr Memorial Arboretum. This complex includes self-guiding trails and a 8,026-acre botanical area.

Continuing westward again on the Byway, you will have ample opportunity for roadside botanizing and valley viewing from overlooks. Talimena State Park (where camping is available) marks the end of the scenic byway and the westward extent of the Ouachita National Forest. The Talihina Visitor Center offers books and maps interpreting the region, as does the Ranger District Office in the town of Talihina.

SALLISAW CREEK STATE PARK

Although a smaller area than the Ouachitas of Oklahoma, the Ozark Plateau of the state supports an interesting mixture of oak-hickory forests, woodlands, savannas, and scattered prairies. Before European settlement the Osage and Caddo tribes occupied this region; Cherokees and other tribes were displaced here from the East during the 19th century.

Just off I-40 in southcentral Oklahoma is the town of Sallisaw; west a few miles via Hwy. 64 is Sallisaw Creek State Park. In the park and environs you can view flora typical of the southwestern Ozark Plateau. Lodging and other services are available in Sallisaw. Robert S. Kerr Lake, 8 miles south, is also recommended for botanizing. Sequoyah's Home Site, a short distance northwest of Sallisaw, includes Sequoyah's log cabin and a visitors' center in a 10-acre wooded area. A Cherokee born in the Appalachians of Tennessee, Sequoyah is recognized as the inventor of the Cherokee nation's writing system.

LAKE TENKILLER, TENKILLER STATE PARK, AND COOKSON HILLS

From Sallisaw take Hwy. 64 west to Gore and then Hwy. 100 north to Lake Tenkiller. Island View Nature Trail begins near the overlook at the south end of the lake. Across the dam is Tenkiller State Park, a popular and often crowded park with cabins and camping facilities. North on Hwy. 100, past the intersection of Hwy. 82 and to the right of the road, is Cookson Hills. A rugged, undeveloped area, it offers ample opportunities for plant studies.

Bibliography

Abrams, M. D. 1992. Fire and the development of oak forests. *BioScience* 42:346–53.

Boon, Bill. 1990. *Nature's Heartland: Native Plant Communities of the Great Plains.* Ames: Iowa State University Press.

Braun, E. Lucy. 1950. *Deciduous Forests of Eastern North America.* New York: Hafner Publishing.

Case, Fredrick W., and Roberta B. Case. 1997. *Trilliums.* Portland, Ore.: Timber Press.

Chapman, Carl H., and Eleanor F. Chapman. 1983. *Indians and Archeology of Missouri.* Columbia: University of Missouri Press.

Clark, G.T. 1977. Forest communities of Crowley's Ridge. *Proc. Arkansas Academy of Science* 28:31–34.

Dale, Edward E., Jr., and Stewart Ware. 1999. Analysis of oak-hickory-pine forests of Hot Springs National Park in the Ouachita Mountains, Arkansas. *Castanea* 64: 163–74.

Daniels, Martha, and Charlotte Overby, eds. 1995. *Missouri Nature Viewing Guide.* Jefferson City: Missouri Department of Conservation.

Denison, Edgar. 1998. *Missouri Wildflowers.* 5th ed. Helena, Mont.: Falcon Press.

Duncan, Wilbur H., and Marion B. Duncan. 1999. *Wildflowers of the Eastern United States.* Athens: University of Georgia Press.

Dungan, Patrick, ed. 1993. *Wetlands.* New York: Oxford University Press.

Elias, Thomas S., and Peter A. Dykeman. 1982. *Field Guide to North American Edible Wild Plants.* New York: Outdoor Life Books.

Flora of North America Editorial Committee. 1997. *Flora of North America North of Mexico.* Vol. 3. Oxford: Oxford University Press.

Foster, Steven, and Roger Caras. 1994. *A Field Guide to Venomous Animals and Poisonous Plants.* Boston: Houghton Mifflin.

Foster, Steven, and James A. Duke. 2000. *A Field Guide to Medicinal Plants and Herbs.* Boston: Houghton Mifflin.

Hauser, Susan C. 1996. *Nature's Revenge.* New York: Lyons and Burford.

Hawkins, T. K., and E. L. Richards. 1995. A floristic study of two bogs on Crowley's Ridge in Greene County, Arkansas. *Castanea* 60:233–44.

Hemmerly, Thomas E. 1990. *Wildflowers of the Central South.* Nashville: Vanderbilt University Press.

———. 2000. *Appalachian Wildflowers.* Athens: University of Georgia Press.

Hunter, Carl G. 1984. *Wildflowers of Arkansas*. Little Rock: Ozark Society Foundation.

———. 1989. *Trees, Shrubs, and Vines of Arkansas*. Little Rock: Ozark Society Foundation.

Kindscher, Kelly. 1992. *Medicinal Wild Plants of the Prairie: An Ethnobotancial Guide*. Lawrence: University of Kansas Press.

Kral, Robert, and Vernon Bates. 1991. A new species of *Hydrophyllum* from the Ouachita Mountains of Arkansas. *Novon* 1:60–66.

Kricher, John C., and Gordon Morrison. 1988. *A Field Guide to Eastern Forests*. Boston: Houghton Mifflin.

Küchler, A. W. 1964. *Potential Natural Vegetation of the Conterminous United States*. Amer. Geogr. Spec. Publ. No. 36.

Kurz, Don. 1996. *Scenic Driving the Ozarks, including the Ouachita Mountains*. Helena, Mont.: Falcon Press.

———. 1999. *Ozark Wildflowers*. Helena, Mont.: Falcon Press.

Ladd, Doug. 1995. *Tallgrass Prairie Wildflowers*. Helena, Mont.: Falcon Press.

Lawton, Barbara P. 1994. *Seasonal Guide to the Natural Year: A Month by Month Guide to Natural Events—Illinois, Missouri, Arkansas*. Golden, Colo.: Fulcrum Publishing.

Leake, Henderson, and Dorothy Leake. 1981. *Wildflowers of the Ozarks*. Little Rock: Ozark Society Foundation.

Little, Elbert L., Jr. 1996. *Forest Trees of Oklahoma*. Oklahoma City: Oklahoma Forestry Service, State Department of Agriculture.

Liggio, Joe, and Ann O. Liggio. 1999. *Wild Orchids of Texas*. Austin: University of Texas Press.

Lyon, John G. 1993. *Practical Handbook for Wetland Identification and Delineation*. Boca Raton, Fla.: Lewis Publishers.

Martin, W. H. 1992. Characteristics of old growth mixed-mesophytic forests. *Nat. Areas J.* 12:129–35.

McFall, Don, ed. 1991. *A Directory of Illinois Nature Preserves*. Springfield, Ill.: Division of Natural Heritage.

McKinney, Landon E. 1992. *A Taxonomic Revision of the Acaulescent Blue Violets* (Viola) *of North America*. Frankfort: Kentucky State Nature Preserves Commission.

McPherson, Alan. 1993. *Fifty Nature Walks in Southern Illinois*. Vienna, Ill.: Cache River Press.

Miller, George O. 1995. *The Ozarks: The People, the Mountains, the Magic*. Stillwater, Minn.: Voyageur Press.

Missouri Natural Areas Committee. 1996. *Directory of Missouri Natural Areas*. Jefferson City: Missouri Department of Conservation.

Mohlenbrock, Robert H., and John W. Thieret. 1987. *Trees: A Quick Reference Guide to Trees of North America*. New York: Macmillan.

Mohlenbrock, Robert H., and John W. Voigt. 1959. *A Flora of Southern Illinois*. Carbondale: Southern Illinois University Press.

Peterson, Lee. 1978. *A Field Guide to Edible Wild Plants*. Boston: Houghton Mifflin.

Peterson, Roger Tory, and Margaret McKenney. 1968. *A Field Guide to Wildflowers*. Boston: Houghton Mifflin.

Petrides, George A. 1998. A Field Guide to Eastern Trees. Boston: Houghton Mifflin.

Phillips, Jan. 1995. *Wild Edibles of Missouri*. Jefferson City: Conservation Commission of the State of Missouri.

Rafferty, Milton D., and John C. Catau. 1991. *The Ouachita Mountains*. Norman: University of Oklahoma Press.

Roberts, David C. 1996. *A Field Guide to Geology: Eastern North America*. Boston: Houghton Mifflin.

Seymour, Randy. 1997. *Wildflowers of Mammoth Cave National Park*. Lexington: University of Kentucky Press.

Smith, Edwin B. 1988. *An Atlas and Annotated List of the Vascular Plants of Arkansas*. 2d ed. Fayetteville: University of Arkansas Bookstore.

———. 1994. *Keys to the Flora of Arkansas*. Fayetteville: University of Arkansas Press.

Smith, Richard M. 1998. *Wildflowers of the Southern Mountains*. Knoxville: University of Tennessee Press.

Steyermark, Julian A. 1963. *Flora of Missouri*. Ames: Iowa State University Press.

Summers, Bill. 1987. *Missouri Orchids*. Jefferson City: Conservation Commission of the State of Missouri.

Turner, Nancy J., and Adam F. Szczawinski. 1991. *Common Poisonous Plants and Mushrooms of North America*. Portland, Ore.: Timber Press.

Vogel, Virgil J. 1970. *American Indian Medicine*. Norman: University of Oklahoma Press.

Williams, John G., and Andrew E. Williams. 1983. *Field Guide to Orchids of North America*. New York: Universe Books.

Yatskievych, George. 1999. *Steyermark's Flora of Missouri*. Vol. 1. St. Louis: Missouri Botanical Garden Press.

Yatskievych, George, and Joanna Turner. 1990. *Catalogue of the Flora of Missouri*. St. Louis: Missouri Botanical Garden.

Index